2nd International Conference on Measurement Instrumentation and Electronics (ICMIE 2017)

AA005651

Journal of Physics: Conference Series Volume 870

Prague, Czech Republic
9 – 11 June 2017

ISBN: 978-1-5108-4444-5
ISSN: 1742-6588

Printed from e-media with permission by:

Curran Associates, Inc.
57 Morehouse Lane
Red Hook, NY 12571

Some format issues inherent in the e-media version may also appear in this print version.

Printed by Curran Associates, Inc. (2017)

Additional copies of this publication are available from:

Curran Associates, Inc.
57 Morehouse Lane
Red Hook, NY 12571 USA
Phone: 845-758-0400
Fax: 845-758-2633
Email: curran@proceedings.com
Web: www.proceedings.com

Table of contents

Volume 870

2nd International Conference on Measurement Instrumentation and Electronics
9–11 June 2017, Prague, Czech Republic

Accepted papers received: 30 June 2017
Published online: 11 July 2017

Preface

011001
OPEN ACCESS
2nd International Conference on Measurement Instrumentation and Electronics

011002
OPEN ACCESS
2nd International Conference on Measurement Instrumentation and Electronics

011003
OPEN ACCESS
Peer review statement

Computer and communication system

012001
OPEN ACCESS
Development of Thread-compatible Open Source Stack

Lukas Zimmermann, Nidhal Mars, Manuel Schappacher and Axel Sikora.....1

012002
OPEN ACCESS
Legacy to Industry 4.0: A Profibus Sniffer

Fesseha Tsegaye Mamo, Axel Sikora and Christoph Rathfelder.....10

012003
OPEN ACCESS
Experimental Identification and Characterization of Multirotor UAV Propulsion

Denis Kotarski, Matija Krznar, Petar Piljek and Nikola Simunic.....16

012004
OPEN ACCESS
Detection of grapes in natural environment using HOG features in low resolution images

Pavel Škrabánek and Filip Majerík.....25

012005
OPEN ACCESS
Impact of corpus domain for sentiment classification: An evaluation study using supervised machine learning techniques

Redouane Karsi, Mounia Zaim and Jamila El Alami.....33

012006
OPEN ACCESS
Dynamic mapping of EDDL device descriptions to OPC UA

Kofi Atta Nsiah, Manuel Schappacher and Axel Sikora.....40

System design & measurement and control technology

012007
OPEN ACCESS
Impact of Machine Virtualization on Timing Precision for Performance-critical Tasks

Kirill Karpov, Irina Fedotova and Eduard Siemens.....48

012008
OPEN ACCESS
The detection of dissolved gases in transformer oil by gas chromatography with helium ionization detector

Xian-qin Deng, Hua Fang and Min-xian Li.....57

012009
OPEN ACCESS
A method of reconstructing the spatial measurement network by mobile measurement transmitter for shipbuilding

Siyang Guo, Jiarui Lin, Linghui Yang, Yongjie Ren and Yin Guo.....64

012010
OPEN ACCESS
DMI measurements impact on a position estimation with lack of GNSS signals during Mobile Mapping

K Bobkowka, G Nykiel and P Tysiąc.....73

Power electronics and electrical engineering

012011
OPEN ACCESS
WAMS measurements pre-processing for detecting low-frequency oscillations in power systems

P Y Kovalenko.....79

012012
OPEN ACCESS
Defining the parameters of a power transmission line equivalent circuit on the basis of phasor measurements

P Y Kovalenko and A S Berdin.....84

012013
OPEN ACCESS
Comparing the techniques of defining the synchronous machine load angle

P Y Kovalenko and A N Moiseichenkov.....91

012014
OPEN ACCESS
Influence of measurement levels number on the accuracy of calculated estimate during the electrical energy measurements verification

E S Kochneva and A Sukalo.....98

012015
OPEN ACCESS
Testing equations method for electrical energy measurements systematic errors detection and measurements results adjustment

E S Kochneva, A V Pazderin and A Sukalo.....105

012016
OPEN ACCESS
Implementation in an FPGA circuit of Edge detection algorithm based on the Discrete Wavelet Transforms

Issam Bouganssa, Mohamed Sbihi and Mounia Zaim.....112

012017
OPEN ACCESS
Piezoelectric actuators in the active vibration control system of journal bearings

J Tůma, J Šimek, M Mahdal, M Pawlenka and R Wagnerova.....121

Materials science and Engineering

012018
OPEN ACCESS
Experimental research on durability of recycled aggregate concrete under freeze- thaw cycles

Yanqiu Cheng, Xiaoyu Shang and Youjia Zhang.....127

012019
OPEN ACCESS
Experimental Study on Abrasive Water Jet Machining of PZT Ceramic

Ajit Dhanawade, Ravi Upadhyai, Arunkumar Rouniyar and Shailendra Kumar.....131

Power machinery and equipment maintenance

012020
OPEN ACCESS
Research on Attribute Reduction in Hoisting Motor State Recognition of Quayside Container Crane

F Li, G Tang and X Hu.....143

012021
OPEN ACCESS
Finite Element Analysis of Doorframe Structure of Single Oblique Pole Type in Container Crane

X. F. Cheng, F. Q. Wu, G. Tang and X. Hu.....148

012022
OPEN ACCESS
State Recognition and Visualization of Hoisting Motor of Quayside Container Crane Based on SOFM

Z.Q. Yang, P. He, G. Tang and X. Hu.....154

012023
OPEN ACCESS
The Effect of the Khong Wong Yai Parameters on Sounds by FEM

Joompon Bamrungwong and Pattaraweerin Woraratsoontorn.....159

Architectural design and project management

012024
OPEN ACCESS
Risk Assessment on Constructors during Over-water Riprap Based on Entropy Weight and FAHP

Tongqing Wu, Liang Li, Zelong Liang, Tian Mao and Weifeng Shao.....166

012025
OPEN ACCESS
Calculation of Appropriate Minimum Size of Isolation Rooms based on Questionnaire Survey of Experts and Analysis on Conditions of Isolation Room Use

An-Na Won, Hae-Eun Song, Young-Kwon Yang, Jin-Chul Park and Jung-Ha Hwang.....173

Environmental analysis and detection

012026
OPEN ACCESS
Research on fluorescence detection method of Microcystis aeruginosa

Xiao-xiong Wang.....178

IOP Conf. Series: Journal of Physics: Conf. Series **870** (2017) 011001 doi:10.1088/1742-6596/870/1/011001

PREFACE

It is our great pleasure to welcome you to 2017 2nd International Conference on Measurement Instrumentation and Electronics which has been held in Prague, Czech Republic during June 9-11, 2017. ICMIE 2017 is dedicated to issues related to measurement instrumentation and electronics.

The major goal and feature of the conference is to bring academic scientists, engineers, industry researchers together to exchange and share their experiences and research results, and discuss the practical challenges encountered and the solutions adopted. Professors from Czech Republic, Germany and Italy are invited to deliver keynote speeches regarding latest information in their respective expertise areas. It is a golden opportunity for the students, researchers and engineers to interact with the experts and specialists to get their advice or consultation on technical matters, teaching methods and strategies.

These proceedings present a selection from papers submitted to the conference from universities, research institutes and industries. All of the papers were subjected to peer-review by conference committee members and international reviewers. The papers selected depended on their quality and their relevancy to the conference. The volume tends to present to the readers the recent advances in the field of computer and communication system, system design and measurement and control technology, power electronics and electrical engineering, materials science and engineering, power machinery and equipment maintenance, architectural design and project management, environmental analysis and detection etc.

We would like to thank all the authors who have contributed to this volume and also to the organizing committee, reviewers, speakers, chairpersons, and all the conference participants for their support to ICMIE 2017.

ICMIE 2017
Organizing Committee
June 20th, 2017

IOP Conf. Series: Journal of Physics: Conf. Series **870** (2017) 011002 doi:10.1088/1742-6596/870/1/011002

Committees

Conference Chairs

Prof. Axel Sikora, Offenburg University of Applied Sciences, Germany
Prof. Ben Choi, Computer Science Louisiana Tech University, USA

Local Chairs

Prof. Vit Vozenilek, Palacky University, Czech Republic
Dr. Pavel Kucera, Eaton European Innovation Center (EEIC) in Prague, Czech Republic

Technical Program Committee Chair

Prof. Wang Shuhui, Tongji University, China
Prof. Ing. Jiří Tůma, Technical University of Ostrava, Czech Republic
Prof. Mohamed Sbihi, Mohammed V University in Rabat Morocco, Morocco
Prof. Natcha Mahapoonyanont, Thaksin University, Thailand

Publication Chair

Asst. Prof. Jakub Svatoš, Czech Technical University, Czech Republic

Technical Program Committee

Dr. Mattia Butta, Czech Technical University in Prague, Czech Republic
Dr. Ilona Janáková, Brno University of Technology, Czech Republic
Dr. Martin STAVA, Brno University of Technology, Czech Republic
Asst. Prof. Jakub Svatoš, Czech Technical University, Czech Republic
Assoc. Prof. Lamia Youb, University of Batna, Algeria
Asst. Prof. Paulo Gil, University of Coimbra, Portugal
Assoc. Prof. Vladimir Shurenkov, National Research Nuclear University, Russia
Ing. Petr Doležel, University of Pardubice, Czech Republic
Dr. Masood Ur Rehman, University of London, UK
Prof. Koichi Takiguchi, Ritsumeikan University, Japan
Prof. Haldun M. Ozaktas, Bilkent University, Turkey
Asst. Prof. Prof. Veysel Gokhan Bocekci, marmara university, Turkey
Asst. Prof. Prof. Kazim Yildiz, Marmara University, Turkey
Prof. Khaled Nagaty, British University in Egypt
Prof. Mounia Zaim, Mohammed V University in Rabat Morocco, Morocco
Dr. Hsien-Shun Liao, National Taiwan University, Taiwan
Prof. Eduard Siemens, The Anhalt University of Applied Sciences, Germany
Asst. Prof. Dr. Parveen Beg, Aligarh Muslim University, India
Mr. Denis Kotarski, Karlovac University of Applies Sciences, Croatia

IOP Conf. Series: Journal of Physics: Conf. Series **870** (2017) 011003 doi:10.1088/1742-6596/870/1/011003

Peer review statement

All papers published in this volume of *Journal of Physics: Conference Series* have been peer reviewed through processes administered by the proceedings Editors. Reviews were conducted by expert referees to the professional and scientific standards expected of a proceedings journal published by IOP Publishing.

IOP Conf. Series: Journal of Physics: Conf. Series **870** (2017) 012001 doi:10.1088/1742-6596/870/1/012001

Development of Thread-compatible Open Source Stack

Lukas Zimmermann, Nidhal Mars, Manuel Schappacher and Axel Sikora

Institute of Reliable Embedded Systems and Communication Electronics, University of Applied Sciences Offenburg, D77652 Offenburg, Germany

lukas.zimmermann@hs-offenburg.de

Abstract. The Thread protocol is a recent development based on 6LoWPAN (IPv6 over IEEE 802.15.4), but with extensions regarding a more media independent approach, which – additionally – also promises true interoperability. To evaluate and analyse the operation of a Thread network a given open source 6LoWPAN stack for embedded devices (emb::6) has been extended in order to comply with the Thread specification. The implementation covers Mesh Link Establishment (MLE) and network layer functionality as well as 6LoWPAN mesh under routing mechanism based on MAC short addresses. The development has been verified on a virtualization platform and allows dynamical establishment of network topologies based on Thread's partitioning algorithm.

1. Introduction

Nowadays, several so-called megatrends like Internet of Things (IoT) are rapidly emerging and are leading to an increasing amount of new protocols and solutions to cover the manifold requirements of different applications. Wireless technologies based on embedded systems play a major role for the local communication. Since the use of IP-based addressing protocols became a necessity to give IoT devices the access to cloud services and interact together, the major drawback was the limited number of unique addresses provided by the IPv4 protocol. This issue was addressed by the development of IPv6 which allows a nearly unlimited number of devices and therefore provides the basis for the IoT. However, when using the IPv6 protocol for 'things' which are mainly embedded devices, several restrictions regarding the underlying technologies and performance of the devices must be considered. Therefore, the usage of IPv6 for embedded devices requires an adaptation layer named 6LoWPAN (IPv6 Low Power Wireless Personal Area Networks) [1] in order to be used in low power wireless networks.

Nevertheless, even with a common network layer there are still several points which are not defined by 6LoWPAN such as the routing protocol and its parameters what might lead to incompatibilities. Therefore, the Thread protocol defines all the layers and technologies required to create a common communication network. To evaluate such a Thread network and to analyse its operation, we extended an open source 6LoWPAN stack – namely emb::6[a] - in order to comply with the Thread specification for connected home applications. The paper is structured as follows: Section 2 provides an overview of the emb::6 stack and the Thread protocol explaining the basic ideas behind. Sections 3 and 4 give insight into selected Mesh Link Establishment (MLE) and network layer functionality. In section 5 we present the memory usage of our implementation. The last section concludes our work. Thread related content is based on [2].

[a] https://github.com/hso-esk/emb6

2. Related Work

The Thread implementation developed and presented herein is using the 6LoWPAN communication stack emb::6. It mainly covers the layer 2 and layer 3 requirements from the Thread Specification [2].

2.1. emb::6 networking stack

emb::6 [3][4] is an open source wireless communication stack based on the 6LoWPAN communication protocol [8]. The initial development of the emb::6 network stack has started as a fork of Contiki OS[b] with several changes regarding the basic architecture. In general, emb::6 covers the Application Layer, Transport Layer, Network Layer, Data Link Layer, Physical Layer, and the Radio Driver. The basic architecture of the emb::6 wireless network stack consists of its networking core, a Hardware Abstraction Layer (HAL), Radio Drivers (RF) and a separate so-called utility module that implements all common functionalities such as timer and event handling used by all other layers and modules. The networking core handles the network related tasks, mainly the communication part, where different tasks are distributed over several layers. Beginning on top at the Application Layer (APL), usually serving as interface for the device application, the stack forwards requests layer by layer down to the radio driver, which is responsible for the implementation of the RF module drivers.

2.2. Thread protocol

The Thread protocol is an open standard for reliable, cost-effective, low power, wireless device-to-device communication. It is designed specifically for connected home applications where IP-based networking is desired and a variety of application layers can be used on the stack. The Thread standard is based on IEEE 802.15.4 (2006) MAC and physical layer operating at 250 kb/s in the 2.4 GHz band. Figure 1 illustrates a general overview of our Thread stack implementation architecture [6]. This work mainly concentrates on MLE and network layer as described in chapter 4 and 5 in the Thread specification.

2.3. Thread device types

The Thread network uses different types of devices as illustrated in figure 2.

- Border Router: A specific type of router that supports multiple interfaces besides IEEE 802.15.4 in order to connect with other networks, e.g. Wi-Fi, Ethernet, etc.
- Router: Used to provide routing services to the network and handle joining and security services for devices trying to join the network. Routers are not allowed to operate as sleepy end devices, but may downgrade their functionality and become REEDs (Router-eligible End Devices).
- Leader: The device that makes decisions within the Thread network and manages router ID assignments. The Leader is the first active router on the network and can be elected in case of losing connectivity.
- Router-eligible End Devices: REEDs have the capability to become routers without user interaction, if necessary.
- End Devices: End devices communicate only through their parent router and cannot forward messages to other devices. To save energy they can sleep for a time period and poll their associated router for data once they are awake.

[b] https://github.com/contiki-os/contiki

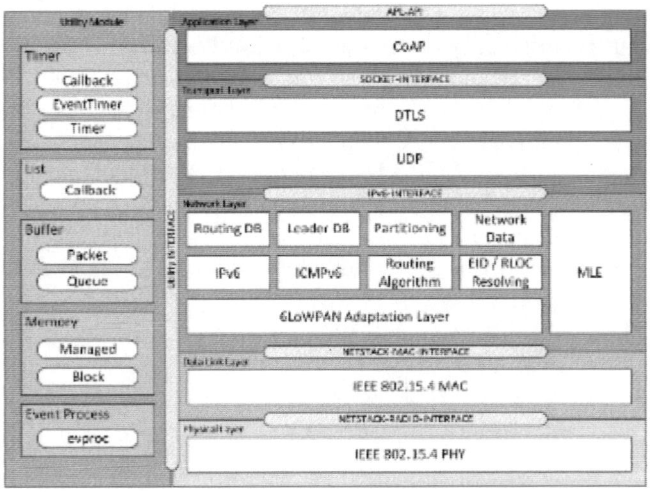

Figure 1. Thread network stack.

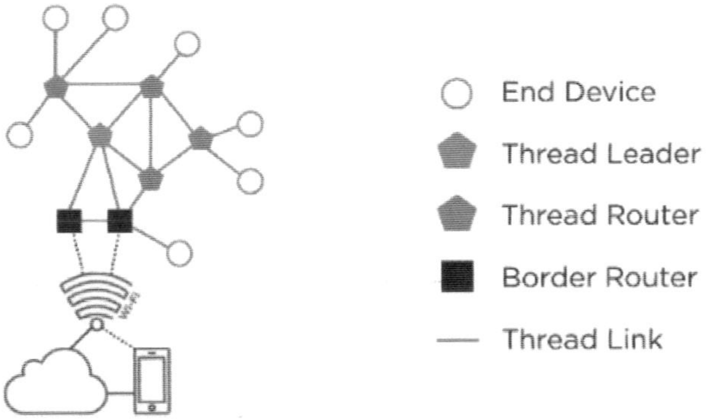

Figure 2. Thread device types [7].

3. Mesh Link Establishment

The existence of many asymmetric radio links within the IEEE 802.15.4 network represents one of the main issues while establishing links between nodes. Thread is using the so-called Mesh Link Establishment protocol (MLE) [8] to resolve such kind of problems in addition to other capabilities.

In this section, we give an overview about the capabilities of the MLE layer and its architecture. Furthermore, we will highlight the main processes of MLE that have been implemented.

3.1. MLE capabilities and architecture

MLE is a protocol that is used to configure and secure radio links dynamically as the topology and physical environment change. This is done by exchanging IEEE 802.15.4 radio parameters between nodes such as addresses, node capabilities and frame counters.

MLE allows all nodes to synchronize periodically and share radio link parameters to adapt to any change that might happen on the topology such as joining of new devices. Furthermore, MLE can detect unreliable links before any effort is spent for configuring them. For example, a link between two devices that is strong in one direction may be unusable due to weak signal strength towards the other direction. MLE resolves this by allowing a node to send periodical link-local multicast messages containing an estimated link quality for all links. In addition, MLE exchanges link costs between nodes by sending MLE advertisement messages.

IOP Conf. Series: Journal of Physics: Conf. Series **870** (2017) 012001 doi:10.1088/1742-6596/870/1/012001

MLE advertisement messages are used to exchange bidirectional link quality between neighboring routers. All routers exchange periodically single-hop MLE advertisement packets containing link cost information. Those periodic advertisements allow routers to quickly detect changes in the set of neighboring routers. For instance, if a new router joins the network, an existing router has been downgraded to REED or if a router lost connection to the Thread network.

Regarding the architectüre, MLE cannot be placed in OSI Model clearly. Instead, it operates alongside the stack using UDP (User Datagram Protocol) as transport protocol. This architecture is also given for other systems that make use of MLE, such as ARM mbed OS [9]. Figure 3 shows the different protocol modules used by ARM mbed OS and the interaction of the MLE protocol with the existing layers.

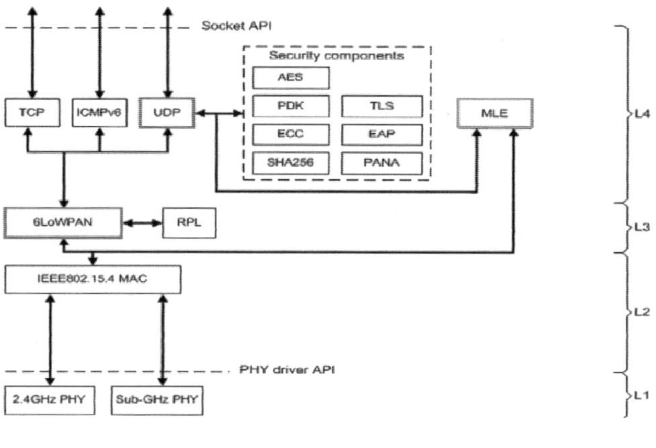

Figure 3. ARM 6LoWPAN stack alongside OSI model [9].

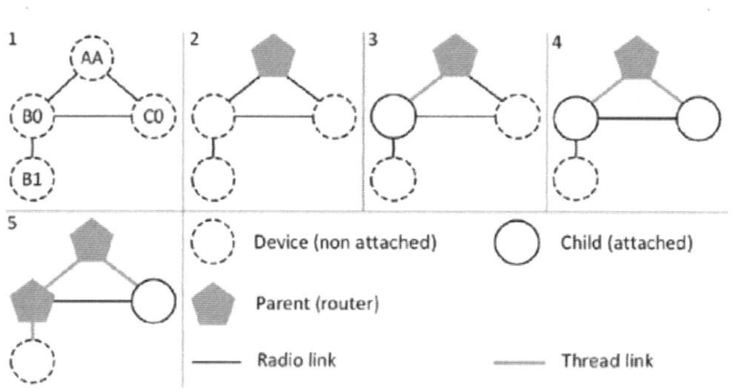

Figure 4. Testing scenario for MLE joining process.

3.2. MLE processes and test cases

In Thread networks, all devices join to the network either as an end device or as a REED. Joiner devices always try to attach to an active Thread router from which they allocate a 16-bit short address. In case such a join process fails, a second request is sent to both routers and REEDs.

Figure 4 briefly shows such a mesh link establishment scenario. We use four nodes that have 0xAA, 0xB0, 0xC0 and 0xB1 as last two bytes of their MAC address, respectively. Possible radio links are defined statically by the environment and are delineated by the gray line.

Figure 5 shows the trace output of the nodes for this scenario. Window A corresponds to node 1 (0xAA) and window B to node 2 (0xB0).

- Node 1 (0xAA) is the first active node in the network. The joining process fails since no other routers are available (line A.12). Consequently, the node creates a new partition and starts operating as a parent (lines A.13 and A.14).
- Node 2 (0xB0) attaches to node 1 after exchanging four handshake messages. Node 2 operates as a child after receiving the CHILD ID RESPONSE (lines B.18 and B.20).
- Node 3 (0xC0) sends a multicast PARENT REQUEST. Node 1 and node 2 receive the message (line A.22 and B.13), but only node 1 replies due to a scan mask TLV (the first request should be replied only by the active router). In case that more than one parent responds, the joining device compares them and selects the best device to be its parent using the connectivity TLV received in the parent response and the calculated two-way link quality (calculated using the received link margin TLV in the parent response and the RSSI of the response itself). Then the handshake process will continue normally between node 3 and node 1. Finally, node 3 (0xC0) will start operating as a child.
- Node 4 (0xB1) has one possible radio link with node 2. Although, node 2 (0xB0) only replied for the second PARENT REQUEST (line B.24). This is explained by the fact that on the first request only the active router should reply (node 2 is operating as a child at that moment). Once node 2 receives a CHILD ID REQUEST, it sends a request to the leader to become a router (line B.26). Finally, node 2 switches its mode to active router (line B.27) and node 4 starts operating as a child.

```
A
1  MLE UDP initialized :
2   lport --> 19788
3   rport --> 19788
4  MLE protocol initialized.
5  [+] JP Send mcast parent request to active router
6  ==> MLE PARENT REQUEST sent to : ff02::2
7  [+] JP Waiting for incoming response from active router
8  [+] JP Send mcast parent request to active Router and REED
9  ==> MLE PARENT REQUEST sent to : ff02::2
10 [+] JP Waiting for incoming response from active Router and
11 REED
12
13 Joining process failed.
14 Starting new partition.
15 MLE : Node operating as Parent.
16
17 [+] SNY process: Send Link Request to neighbor router.
18 ==> MLE LINK REQUEST sent to : ff02::2
19 [+] SNY process:  Synchronization process finished.
20 <== MLE PARENT REQUEST received from : fe80::250:c2ff:fea8:b0
21 ID allocated for the child = 1
22 ==> MLE PARENT RESPONSE sent to : fe80::250:c2ff:fea8:b0
23 <== MLE PARENT REQUEST received from : fe80::250:c2ff:fea8:c0
24 ID allocated for the child = 2
25 ==> MLE PARENT RESPONSE sent to : fe80::250:c2ff:fea8:c0
26 <== MLE CHILD ID REQUEST received from : fe80::250:c2ff:fea8:b0
27 ==> MLE CHILD ID RESPONSE sent to : fe80::250:c2ff:fea8:b0
28 <== MLE CHILD ID REQUEST received from : fe80::250:c2ff:fea8:c0
29 ==> MLE CHILD ID RESPONSE sent to : fe80::250:c2ff:fea8:c0
30 Child linked with id : 1 and timeout is : 10
```

```
B
1 MLE UDP initialized :
2  lport --> 19788
3  rport --> 19788
4 MLE protocol initialized.
5 [+] JP Send mcast parent request to active router
6 ==> MLE PARENT REQUEST sent to : ff02::2
7
8 [+] JP Waiting for incoming response from active
9 router<== MLE PARENT RESPONSE received from :
1 fe80::ff:fe00:0
0 [+] JP received response from active Router
1 my rssi : 30
1 rssi of parent : 30
1 two-way link quality : 3
2 <== MLE PARENT REQUEST received from :
1 fe80::250:c2fffea8:c0
3 [+] JP Parent selection
1 link quality with parent : 3
4 [+] JP send Child ID Request
1 ==> MLE CHILD ID REQUEST sent to : fe80::ff:fe00:0
5 <== MLE CHILD ID RESPONSE received from :
1 fe80::ff:fe00:0
6 [+] JP Parent Stored
1 MLE : Node operating as Child
7 <== MLE PARENT REQUEST received from :
1 fe80::250:c2ff:fea8:b1
8 <== MLE PARENT REQUEST received from :
1 fe80::250:c2ff:fea8:b1
9 ID allocated for the child = 1
2 ==> MLE PARENT RESPONSE sent to : fe80::250:c2ff:fea8:b1
0 <== MLE CHILD ID REQUEST received from :
2 fe80::250:c2ff:fea8:b1
1 Sending request to become a Router
2 MLE : Node operating as Parent.
```

Figure 5. Trace output of Thread nodes.

4. Network Layer

In this section, we provide information regarding the Thread network layer. Each Thread device must at least support the IEEE 802.15.4 (2006) standard but might support additional interfaces. This makes the Thread protocol appear heterogeneous from a PHY layer point of view. Above the MAC layer, Thread turns into a homogeneous protocol supporting IPv6 and IP routing.

4.1. EID / RLOC separation

In a Thread network, each interface must have a routing locator (RLOC) address and an endpoint identifier (EID) address. The RLOC address is based on the router ID and child ID assigned to an interface. If for example one child switches over to another parent, only the RLOC will change, but not

IOP Conf. Series: Journal of Physics: Conf. Series **870** (2017) 012001 doi:10.1088/1742-6596/870/1/012001

the child's identity at all. Thread nodes are arranged in partitions forming closed affiliation areas for participating devices whereas each partition has one node designated as the leader. The leader is responsible for assigning and managing router IDs. A Thread EID is a stable IPv6 address that uniquely identifies a Thread interface within a Thread partition. EIDs are not directly routable, because the Thread routing protocol only exchanges route information for RLOCs. To deliver an IPv6 datagram with an EID as the IPv6 destination address, a Thread device must perform EID-to-RLOC lookup. When attaching to a partition, a node must retrieve an RLOC IPv6 address from a router. The RLOC's 16 least significant bits are called RLOC16 and map router ID and child ID of the node. Routers assign child ID 0. Figure 6 shows the RLOC16 structure.

0	1	2	3	4	5	6	7	8	9	10	11	12	13	14	15
Router ID						R	Child ID								

Figure 6. RLOC16 structure.

A router retrieves his router ID from the partition leader by sending a CoAP address query message. RLOC addresses are only used for communicating control traffic and delivering IPv6 datagrams to their destinations. Since no RLOC address is available when initially sending an address query message, the EID is used. Intermediate nodes must perform EID-to-RLOC lookup in order to forward the packet to the partition leader and vice versa. The child ID is allocated by the parent node and communicated through MLE attachment process.

4.2. Routing algorithm

A Thread network has up to 32 active routers that use next-hop routing for messages based on its routing database. This database includes path cost calculation that is performed by applying distributed Bellman-Ford algorithm (cf. RIPng) [10]. The routing database is a set of neighbor router table (Link Set), routing table (Route Set) and all valid router IDs (Router ID Set). All routers advertise their routing table periodically. The rate at which routing advertisements are sent is determined by an instance of the Trickle algorithm. For routers, in order to keep track of the validity of shared data in the network, an incremental ID sequence number is attached to the routing data. After looking up the shortest path for a route, a router generates the IPv6 RLOC address of the destination router using its router ID.

All tables that are part of the routing database have been implemented by using linked lists. When looking for a routing entry, linked list structures can be used as a mask when iterating trough all list entries. The benefit of this approach is that predefined fields can be accessed easily. Since embedded devices usually underlie memory constraints, we implemented least recently used (LRU) replacement policy for Link Set and Route Set. The last recently accessed item is inserted at the head of the linked list by modifying appropriate pointers. As a result, when transmitting fragmented packets the lookup iteration for subsequent fragments will terminate after the first list element. When inserting a new list element, the last one of the linked list is removed if the number of elements would exceed a defined maximum.

4.3. Link cost determination

The link set stores information about neighboring routers including the measured link margin (RSSI) in dB. Furthermore, the link margin plays a leading role in parent selection during attachment process. The measured one-way link margin may change during runtime due to noise floor or altered environmental conditions. To smooth out short-term volatility, Thread devices must perform exponentially weighted moving average (EWMA) method of the link margins for each neighbor. Equation (1) shows the EWMA calculation where M_{t-1} is the currently stored link margin for a specific neighbor, Y_t is the last recently measured link margin and M_t will be the newly calculated link margin for that neighbor.

$$M_t = \alpha \cdot Y_t + (1 - \alpha) \cdot M_{t-1} \quad (1)$$

To avoid costly floating point computations on the micro- controller, equation (1) has been rewritten to equation (2).

$$M_t = \frac{Y_t + \left(\frac{1}{\alpha} - 1\right) \cdot M_{t-1}}{\frac{1}{\alpha}} \quad (2)$$

The exponential smoothing fraction α (equation (3)) is used as weighting and defined as either $\frac{1}{8}$ or $\frac{1}{16}$ [11].

$$\alpha = \{\alpha \in R \mid 0 \leq \alpha \leq 1\} \quad (3)$$

The preceding transformation results in a bit shift to the left of the numerator around the reciprocal value of α. This allows to exploit the benefits of integer calculations without glaring rounding errors.

4.4. *Routing advertisement*

Distributed routing algorithms reduce node-sided computational costs in terms of shared route data. When using non-distributed algorithms, each node has to expand a graph by incrementally improving path costs. In Thread, MLE advertisements are used from nodes acting as routers for advertising their routing table to neighboring routers. A practicable approach of determining the rate at which advertisements are sent is to deploy a dependency on the change of routing data. Thread uses the Trickle algorithm to generate dynamic and random transmission windows. If the routing entries are stable, the rate is reduced to a minimum. The flowchart in figure 7 shows our implementation of the Trickle algorithm. We use a timer that recalculates its expiration time after a timeout. The limits of the time slots can be defined via C macros. After initializing the Trickle timer, it runs independently from other processes.

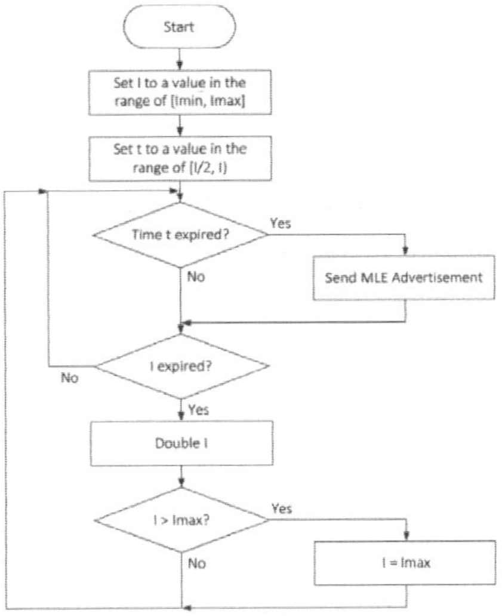

Figure 7. Flowchart of Trickle timer implementation.

4.5. *Unicast packet forwarding*

Routing inside a Thread network is performed using RLOC IPv6 addresses. Unicast packets are forwarded by applying a mesh under strategy on the 6LoWPAN layer. Router packets include the 6LoWPAN mesh header carrying the originator and final RLOC16 addresses. When receiving a packet including a 6LoWPAN mesh header, a routing table lookup is performed without uncompressing the packet. Therefore, we extended the emb::6 implementation in order to support mesh under routing for unicast packets. During MLE joining process, the 16-bit short MAC address is set to the RLOC16 assigned to the router. Usually, IPv6 packets from higher layers, e.g. application layer, are using the EID of the destination device. Then, routers must perform EID-to-RLOC lookup to retrieve the router ID of the destination router. The EID-to-RLOC lookup mechanism consists of CoAP messages targeting CoAP resources provided by routers [12]. The router that is responsible for the given EID is sending a response message including its router ID. Each router maintains a EID-to-RLOC map cache to hold a list of recently used lookups. This prevents from frequently sending lookup messages when transmitting fragmented packets. For instance, an end device does not have routing capability and therefore must forward packets to its parent router. In this case the packet is sent without adding 6LoWPAN mesh header.

5. Memory usage

An analysis of our implementation in terms of memory usage is shown in table 1. Both figures show the RAM and flash memory usage for selected network layer modules and MLE. Mesh under routing is enabled on 6LoWPAN layer. The 'net-thread' bar shows all additional Thread-specific implementations such as routing database, partitioning etc. The reason for the high memory usage of this part is the static memory allocation of the routing database.

Table 1. Thread implementation memory usage.

Feature	RAM (kB)	Flash (kB)
net-ipv6	4,056	25,025
net-sicslowpan	460	14,826
net-thread	12,692	33,014
mle	2,088	15,638

6. Conclusion

This paper presents the architecture and basic information about the Thread protocol. The current status of the implementation allows to create nodes inside a hardware virtualization environment dynamically. Thread nodes autonomously establish connection and create network topologies as specified by Thread algorithms. An important next step will be the full integration of Thread Network Data and commissioning protocol in order to validate the compatibility with existing Thread devices.

References

[1] G Montenegro, N Kushalnagar, J Hui, D Culler, *Transmission of IPv6 Packets over IEEE 802.15.4 Networks*, RFC 4944, (Sept. 2007)

[2] Thread Group, Inc., *Thread Specification*, Revision 1.1.0 (July 2016)

[3] A Yushev, A Sikora and J Sebastian E, *Open source 6Lo protocol stack for wireless embedded systems*, 2016 Wireless Telecommunications Symposium (WTS), London, UK, 2016, pp. 1-7.

[4] M Schappacher, E Schmitt, A Sikora, P Weber, A Yushev, A Flexible, Modular, *Open-Source Implementation of 6LoWPAN*, 8th IEEE International Conference on Intelligent Data Acquisition and Advanced Computing Systems: Technology and Applications (IDAACS2015), 24-26 September 2015, Warsaw, Poland, pp. 838-844.

[5] A Sikora, *Funknetzwerke für das Internet der Dinge: 6LoWPAN OpenSource-Projekt: emb6*, Elektronik Wireless 2016, (2016)

[6] Thread Group, Inc., *Thread Stack Fundamentals (whitepaper)*, Revision 2.0 (July 2015)

[7] B Curtis, S Ashon. *Thread Open House*. Thread Group, (May 2016)

[8] R K Kelsey, *Mesh Link Establishment*, (Oct. 2013)

[9] ARM mbed 6LoWPAN Stack Overview, https://docs.mbed.com/docs/arm-ipv66lowpan-stack/en/latest/02_N_arch/, (04.02.2017)

[10] Malkin, G and R Minnear, *RIPng for IPv6*, RFC 2080, (Jan. 1997)
[11] T Agami Reddy, *Applied data analysis and modeling for energy engineers and scientists*, in. Boston, MA: Springer US, pp. 253–288 (2011)
[12] Z Shelby, K Hartke, C Bormann, *The Constrained Application Protocol (CoAP)*, RFC 7252, (2014)

IOP Conf. Series: Journal of Physics: Conf. Series **870** (2017) 012002 doi:10.1088/1742-6596/870/1/012002

Legacy to Industry 4.0: A Profibus Sniffer

Fesseha Tsegaye Mamo[1, a], Axel Sikora[1,a], Christoph Rathfelder[1,a]

[1]Hahn-Schickard, Software Solutions, 78052 Villingen-Schwenningen, Germany

f.mamo@hahn-schickard.de; christoph.rathfelder@hahn-schickard.de; axel.sikora@hs-offenburg.de

Abstract. Legacy industrial communication protocols are proved robust and functional. During the last decades, the industry has invented completely new or advanced versions of the legacy communication solutions. However, even with the high adoption rate of these new solutions, still the majority industry applications run on legacy, mostly fieldbus related technologies. Profibus is one of those technologies that still keep on growing in the market, albeit a slow in market growth in recent years. A retrofit technology that would enable these technologies to connect to the Internet of Things, utilize the ever growing potential of data analysis, predictive maintenance or cloud-based application, while at the same time not changing a running system is fundamental.

1. Introduction

Legacy industrial communication protocols are proved robust and functional. During the last decades, the industry has invented completely new or advanced versions of the legacy communication solutions. However, even with the high adoption rate of these new solutions, still the majority industry applications run on legacy, mostly fieldbus related technologies. Despite its age, Profibus is one of those technologies which still are very popular in the market. [1] A retrofit technology enabling these technologies to connect to the Internet of Things, would open the ever growing potential of data analysis, predictive maintenance or cloud based applications, while at the same time not changing a running system.

In order to achieve the above described technology, it is necessary to adapt existing communication protocols which are not capable of seamless interconnection to IP based networks. Profibus is one of these protocols which rely on RS485 physical medium for majority of its use cases [2]. In this paper we describe a technique to demonstrate such a possibility of physical and logical interconnections. The demonstrator focuses on a Profibus sniffer that passively collects data from the communication bus, without interfering with the system, and that forwards relevant data to a server for storage and further analysis and manipulation of the collected data.

This paper initially takes a look at comparable solutions and the Profibus protocol. In later sections the specific setup, hardware and software solutions from this work are described in brief.

2. Project Background

The objective of the presented work is part of a larger project. The project named in short NIKI4.0 (Nicht-disruptives Kit für die Evaluation von Industrie 4.0) [3] is a cooperative project between Hahn-Schickard [4], Institute of reliable embedded systems and communication (IvESK) [5] at the University of Applied Sciences Offenburg, and Karlsruhe Forschungszentrum Informatik (FZI) [6]. Many industrial partners also take part in the project. The project is funded by the Baden-Württemberg-Stiftung [7].

The project aims at equipping industries with a retrofit solution by fitting production sites with different kind of sensors, information monitoring tools (Profibus sniffer), and extended visualization

[a] f.mamo@hahn-schickard.de, axel.sikora@hahn-schickard.de, christoph.rathfelder@hahn-schickard.de

mechanisms. With such a system in place it will be possible to collect valuable information that helps in better understanding and optimizing the operations of the facilities.

3. State of the Art

There currently exist few Profibus protocol sniffers. These sniffers focus on capturing raw Profibus data and analyzing the electrical behaviors of the network bus for troubleshooting purposes. Few examples are the ProfiTrace [8] and ibaBM-DP [9]. In addition to data sniffing these devices also provide analysis tools for the electrical signal on the network bus.

Using Wireshark [10] for sniffing communication protocols is the general approach. However, there are no solutions for Profibus protocol sniffing that is based on Wireshark.

4. Profibus Overview

Profibus is defined by a collection of IEC standards (cf. Figure 1) and it is one of the most successful and mostly used industrial communication protocols. Profibus operates on different physical mediums addressing the requirements in different industrial environments and use cases. Profibus defines the specification for the Fieldbus Data Link layer, the Application layer, and the Application profiles. Figure 1 summarizes the components of Profibus as represented according to the OSI model.

Profibus User Association (PNO) [11], through the years, has adopted multiple engineering tools that make installation and commissioning of Profibus devices easy. General Station Description (GSD) [12] and Electronic Device Description (EDD) [13] are examples of such solutions that describe the capabilities of a Profibus device in a generic way. They describe the hardware capabilities of a Profibus slave device starting from supported speeds to the number of input outputs on a slave module.

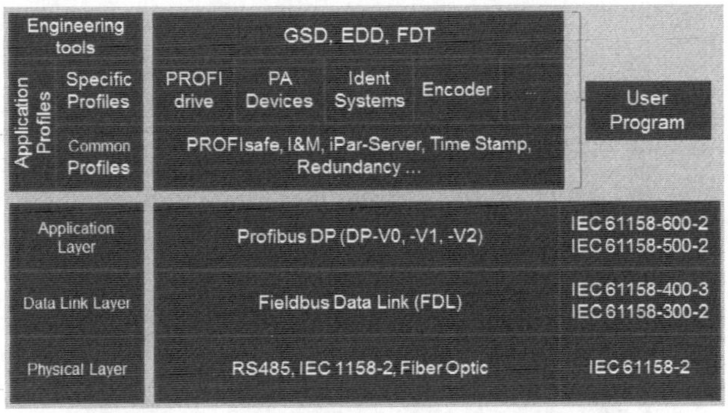

Figure 1. Profibus specification and tools

The Profibus data link layer, named the Fieldbus Datalink Layer (FDL), provides data transmission services that serve different purpose. These services for data transmission are:

- SDN - Send Data with No acknowledge
- SDA - Send Data with Acknowledge
- SRD - Send and Request Data
- CSRD - Cyclic Send and Request Data
- MSRD - Send and Request Data with Multicast Reply
- CS - Clock Synchronization

These transmission services utilize five types of packet structure. These are the telegrams without data (SD1), telegrams with variable length (SD2), telegrams with fixed data (SD3), the token telegram (SD4) and the short confirmation (SC). Actual information with payload can only be transmitted in SD2 and SD3 Fieldbus Datalink Layer (FDL) packets. The structure of these packets is shown below.

SD2=0x68	LE	LEr	SD2	DA	SA	FC	PDU	FCS	ED=0x16

SD3=0xA2	DA	SA	FC	PDU	FCS	ED=0x16

The following list of information can be acquired from a Profibus data bus:

- Diagnosis
- I/O Data
- Alarm
- Process Data
- LR (Load Register) Data
- Context
- Function Invocation Data

Of the above mentioned, diagnosis, alarm, status and I/O data are planned to be retrieved in this research. The GSD file of a Profibus device provides module information which is useful to extract meaningful I/O information from the gathered (sniffed) data.

5. System setup

One of the physical mediums supported by Profibus, as shown in Figure 1, is RS485. Since this is the mostly used physical media this work focuses on Profibus networks using RS485. It is a two wire pair differential system. Profibus slave devices provide a diagnosis port that enables monitoring or diagnosis devices to connect to it, as shown in Figure 2. Using this port and a carefully designed hardware to collect data from this port, all data on the Profibus network can be sniffed. The hardware setup is discussed in later sections (cf. ch. 6.1) of this paper.

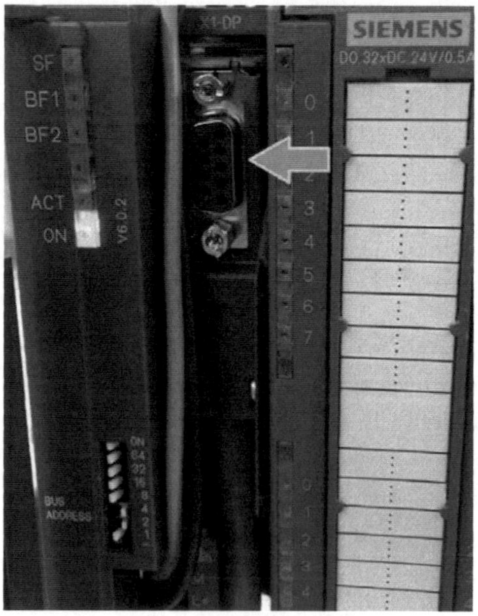

Figure 2. SIMATIC ET200M –IM 163-2 Profibus slave diagnosis port

After the data collection, filtering of relevant traffic is the second step. Relevant traffic in this demonstrator includes diagnosis, alarm, status and I/O data. In this demonstration, once the packets are received at the end of the RS485 interface of the sniffer, the next task is to filter the SD2 and SD3 type packets. After this step modelling of the PDU is needed before transmitting to the gateway. The data is modelled using the IPSO object model; represented using either JSON or XML and the data is transmitted using CoAP which is the transport mechanism in LWM2M [14].

The Profibus sniffer collects data by tapping into the Profibus slave diagnosis port as shown in Figure 3. The data is then provided as a stream for further processing relevant information.

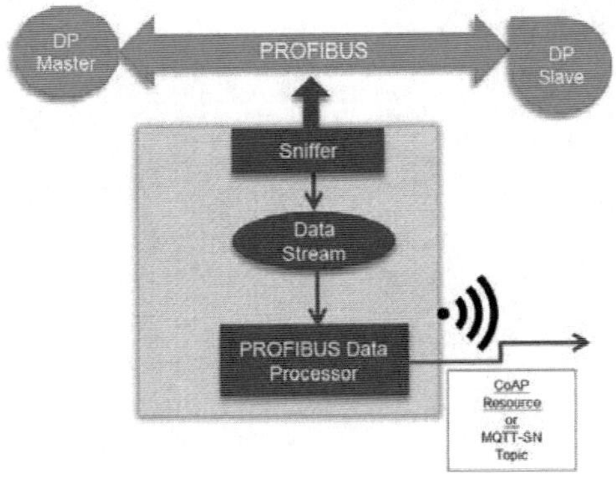

Figure 3. Profibus sniffer

6. Demonstrator

6.1 Hardware setup

Profibus specification supports speeds of up to 12 Mbps. Therefore the RS485 transceiver hardware should be able to operate up to such speed. However normal off the market RS485 to UART converters do not support such speed. In majority of Profibus devices such speeds are possible by using a targeted ASIC designed hardware.

In this research, though, a system with minimum complexity and affordable cost is attained by using the TI Sitara AM3358 processor [15] that provides the programmable real-time core (PRU) for industrial communication protocols, Figure 4. The PRU provides a UART interface with a dedicated clock of 192MHz that can operate on speeds up to 50Mbps.

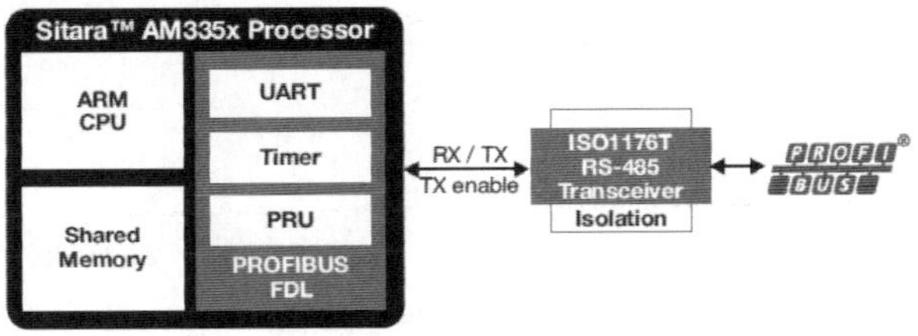

Figure 4. TI Sitara Processor [15]

Such a processor is found in the Beaglebone Black open source hardware development kit. The Profibus sniffer is built by combining the Beaglebone Black's PRU UART port with an external RS485 to UART transceiver module. This setup is shown in Figure 5.

Figure 5. Sniffer Hardware

6.2 Software

DP or Decentralized Peripherals deals with decentralized I/O modules. In simple terms there are an array of input and output modules represented by single bytes and bits, representing a 1 as ON and a 0 as OFF. Different data types are described by the Profibus DP profile.

Using the sniffer hardware, an output of raw data is provided at the Linux subsystem level. This raw Profibus data has to be then fed to a parser to take out the interesting information as described in section 4 of this paper. A sample raw data with description is shown below in Figure 6.

The first packet is a packet with variable data length with start delimiter (SD) 0x68. It is a request packet, identified by the function code (FC=0x21), destined to the destination address of 3, which in this setup is a Slave device. The sender is a Master device with the address 2. The master is requesting IO data for the slot 0x0B and index of 0x04. The length of the requested IO data is signified by the following zero bytes. The 0x91 byte is the Frame Check Sequence and the last byte 0x16 is the fixed value End Delimiter field.

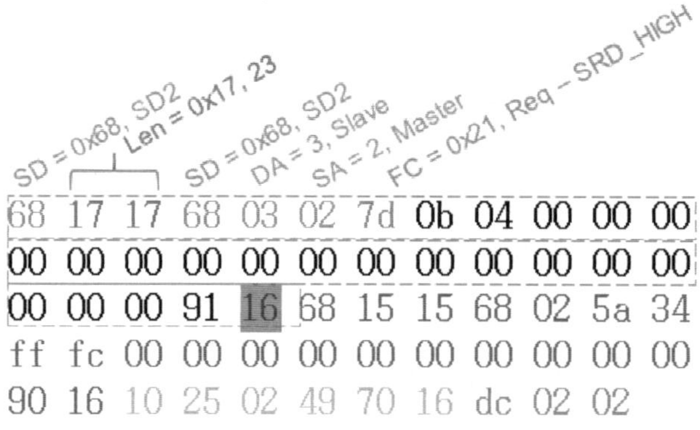

Figure 6. Raw Profibus data

In the case of a distributed object structure for Profibus, the instance should act as the address of the Profibus device.

e.g. /3500/2/6001 represents diagnosis information from the Profibus device with address 2

If the device in the above example reports multiple diagnosis information, a rolling amount of instances of the diagnosis data should be stored. There should be a limit to the stored instance amount in order not starve out memory.

The first approach of representing Profibus data is in a modular, functional oriented way. This means creating a separate object representation based on the type of information contained. The second type of object representation is a device based approach.

IOP Conf. Series: Journal of Physics: Conf. Series **870** (2017) 012002 doi:10.1088/1742-6596/870/1/012002

Table 1. Profibus diagnosis data according to IPSO resource model

Object	Object ID	Object URN	multiple instances?
Profibus Diagnosis	3500		Yes

res name	res ID	access type	multiple instances	M/O	type
dev add	6000	R	No	M	Float
ident num	6001	R	No	M	Float
diag status	6002	R	Yes	M	String
identifier-based diag.	6003	R	Yes	O	String
channel-based diag.	6004	R	Yes	O	String
device-based diag.	6005	R	Yes	O	String

7. Summary

This paper and the work done under this demonstration are a proof of concept for integrating legacy communication protocols to the ever growing and demanding necessity of connecting things to the Internet.

While connecting legacy systems to the Internet seems a fairly uncomplicated task, seamlessly mapping the semantics of such systems is a very challenging task. This is so because of the simplified roles of such kind of field devices and that the actual logic of such systems runs on external systems such as PLCs. To give an example, a Profibus slave device could have multiple digital inputs and outputs which are bit wise programmed for a given functionality on the PLC. Therefore even if it is possible to get these process values by tapping on the network bus, it is impossible to understand the meaning of it unless the context of each bits and bytes are known from the PLC.

With that in mind, it is still possible to collect standard information being described by the Profibus specifications, which are context independent such as diagnosis, alert and status information.

References

1. PI Figures Specification 2014, http://www.profibus.com/nc/pi-organization/regional-pi-associations/south-east-asia/downloads-a nd-media/downloads/2015-pi-seminar/download/19049/
2. http://www.smar.com/en/profibus
3. http://fzi-forschungszentrum-informatik.github.io/NIKI40/
4. http://www.hahn-schickard.de
5. http://ivesk.hs-offenburg.de
6. https://www.fzi.de/startseite
7. https://www.bwstiftung.de/startseite
8. http://www.procentec.de/profitrace2/
9. http://www.iba-ag.com/de/produkte/produktkatalog/13121001-ibabm-dp/
10. https://www.wireshark.org
11. http://www.profibus.com/pi-organization/about-pi/
12. Profibus GSD files, http://www.profibus.com/products/gsd-files/
13. EDDL Specification; Specification for PROFIBUS Device Description, vol. 2
14. Open Mobile Alliance Lightweight M2M, http://www.openmobilealliance.org/wp/Overviews/lightweightm2m_overview.html
15. Profibus on Sitara Processors whitepaper, http://www.ti.com/lit/wp/spry155b/spry155b.pdf

IOP Conf. Series: Journal of Physics: Conf. Series **870** (2017) 012003 doi:10.1088/1742-6596/870/1/012003

Experimental Identification and Characterization of Multirotor UAV Propulsion

Denis Kotarski[1], Matija Krznar[2], Petar Piljek[3], Nikola Simunic[1]

[1] Karlovac University of Applied Sciences, Mechanical Engineering Department, J.J. Strossmayera 9, Croatia
[2] Peti Brod, Zlatarska 14, Croatia
[3] University of Zagreb, Faculty of Mechanical Engineering and Naval Architecture, Ivana Lucica 5, Croatia

denis.kotarski@vuka.hr

Abstract. In this paper, an experimental procedure for the identification and characterization of multirotor Unmanned Aerial Vehicle (UAV) propulsion is presented. Propulsion configuration needs to be defined precisely in order to achieve required flight performance. Based on the accurate dynamic model and empirical measurements of multirotor propulsion physical parameters, it is possible to design diverse configurations with different characteristics for various purposes. As a case study, we investigated design considerations for a micro indoor multirotor which is suitable for control algorithm implementation in structured environment. It consists of open source autopilot, sensors for indoor flight, "take off the shelf" propulsion components and frame. The series of experiments were conducted to show the process of parameters identification and the procedure for analysis and propulsion characterization. Additionally, we explore battery performance in terms of mass and specific energy. Experimental results show identified and estimated propulsion parameters through which blade element theory is verified.

1. Introduction

The ability to act in three-dimensional environment leads to new research challenges in the field of design, modelling and control of UAV [1]. Development of new technologies has enabled rapid progress and a large number of interesting applications. UAVs can be potentially used for accident site monitoring or search and rescue missions [2, 3]. There are several categories of UAVs which are in various stages of research, development and utilization (rotary-wing, fixed wing, flapping-wing, blimp, hybrid UAV). Because of their unique ability to perform vertical take-off and landing, stationary flight and flight at moderate speed, multirotor UAVs will be explored. They represent fixed pitch rotary-wing aircrafts. There are a variety of designs and sizes ranging from large aircraft to the micro air vehicles (MAV). From a mechanical point of view, multirotor has six degrees of freedom (6 DOF). The only moving parts are propellers with constant pitch connected to the rotor axis. It is assumed that the multirotor frame is symmetrical and rigid, thus rotors angular velocities are the only variables that have a direct impact on the multirotor dynamics. [4].

Design and development of multirotors is constrained by the size, weight and power consumption of the propulsion components [5]. Multirotor dynamics depends on the aircraft design and propulsion

IOP Conf. Series: Journal of Physics: Conf. Series **870** (2017) 012003 doi:10.1088/1742-6596/870/1/012003

aerodynamic forces. Batteries, electronic speed controllers (ESC), brushless DC (BLDC) motors and propellers have to be chosen as precisely in order to achieve optimal flight performance. In papers [6,7], the experimental setup for identification of propulsion physical parameters is described. Series of experiments was conducted in [8, 9] to evaluate the efficiency of coaxial propulsion configuration. The paper [10] investigate the effect of interactions between the rotor flow and thrust in order to achieve increase in thrust for a limited body size. Furthermore, authors in [11] examine the influence of propulsion configuration and arrangement on a system efficiency in hover mode. Since the batteries represent a significant constraint in the multirotor design, papers [12,13] investigated the problems of endurance and length of the flight.

In this paper, experimental test bench and procedure for propulsion characterization of the multirotor UAV are presented. Experimental measurements where conducted on components whose selection is based on the proposed case study. Experimental identification, analysis and propulsion characterization enable successful design of a different multirotor configurations. To develop faithful model, it is important to measure parameters which are associated with multirotor dynamics and parameters, important from the standpoint of energy consumption. Test bench design has the ability to test a large range of propulsions (from 2" to 20") with different configurations.
The paper is organized as follows. Section 2 describe design considerations for multirotor MAV. Section 3 presents experimental setup. In section 4, experimental results are shown. Conclusion is discussed in section 5.

2. Design Considerations for MAV

Multirotor design can be classified into two main categories: flat and non-flat configurations. From the control point of view, multirotor is highly nonlinear, multivariable and inherently unstable 6 DOF rigid body.

The conventional flat designs share properties of underactuated and strongly coupled systems. Flat designs are divided with respect to the propulsion configuration. The most common designs incorporate a single propulsion units (quadrotor, hexarotor and octarotor) or coaxial propulsion units (Y6 and X8 rotor) as shown in figure 1. There are also designs with overlapping propulsion arrangements [10].

Non-flat designs with tilted rotors have the ability to overcome the inherent underactuated property of flat designs. In this case it is necessary that multirotor consists of six or more propulsion units. Fully actuated multirotors can decouple position from orientation which has significant influence on the multirotor control design.

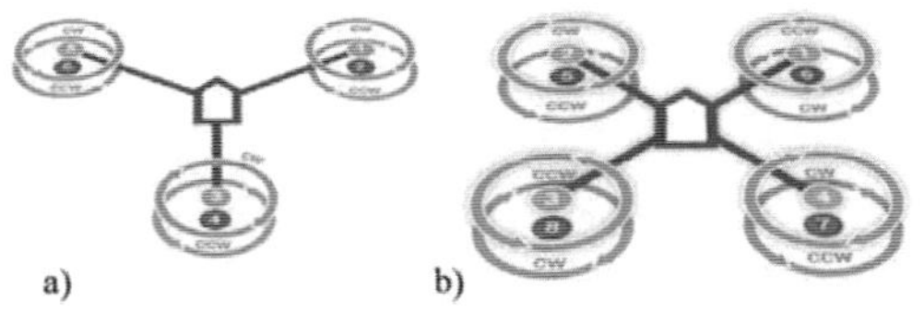

Figure 1. Coaxial flat configurations: a) Y6 rotor, b) X8 rotor (Pixhawk™ motor layout).

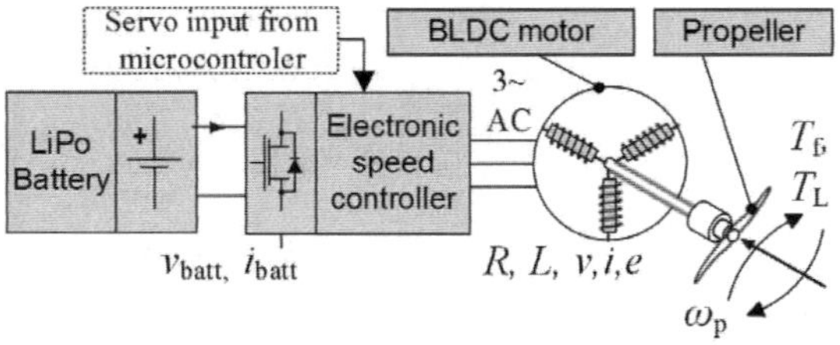

Figure 2. Model of the multirotor propulsion unit.

2.1. Propulsion components selection

For successful multirotor design, it is necessary to indentify the propulsion physical parameters. Take off the shelf components mostly have insufficient or inaccurate data therefore it is necessary to conduct experimental measurements.

Before the measurement procedure, it is necessary to select propulsion components for which the identification and characterization will be conducted (see figure 2). In this research components are selected based on the case study.

2.1.1. Propeller. During rotation, the propeller produces aerodynamic forces that directly affect the multirotor dynamics. They depend on the angular velocity and the propeller geometrical shape. Thrust force of *i*-th propulsion is equal to:

$$f_i = k_f {\omega_i}^2 \tag{1}$$

where ω_i is angular speed of the *i*-th rotor, and k_f is the thrust force factor which can be calculated as:

$$k_f = C_T \rho A r^2 \tag{2}$$

where C_T is thrust coefficient, ρ is air density, A is the area of propeller disk and r is the propeller radius. Drag moment of *i*-th propulsion is equal to:

$$\tau_i = k_\tau {\omega_i}^2 \tag{3}$$

where k_τ is the drag torque factor:

$$k_\tau = C_P \rho A r^3 \tag{4}$$

where C_P is the power coefficient.

This paper proposes the procedure for propeller testing, with various geometric characteristics (given in Table 1), on the whole range of possible rotor angular velocities. Changing the propeller geometric characteristics result in a change of treated air quantity per rotation. The higher quantity of air has a higher motion resistance and thus increases a drag moment, which requires an increase in motor torque. As a result, motor current and power consumption increases. Therefore, selection of the propeller necessarily affects the choice of motors and batteries. The experiment will yield dependence of physical parameters on propeller geometrical features.

Table 1. Considered geometric characteristics of propellers.

Diameter	Pitch	Blade Number	Bull Nose	Label
3''	3''	2	Yes	3030_2_BN
	3''	3	No	3030_3
	3''	4	Yes	3040_4_BN
	4''	3	No	3040_3
4''	4''	2	No	4040_2
	4.5''	2	No	4045_2
	4.5''	3	No	4045_3
	4.5''	2	Yes	4045_2_BN
	4.5''	3	Yes	4045_3_BN
	4''	4	Yes	4040_4_BN
5''	3''	2	No	5030_2
	3''	3	No	5030_3
	4''	2	No	5040_2
	4''	3	No	5040_3
	4''	4	No	5040_4
	4''	6	No	5040_6
	4.6''	2	Yes	5046_2_BN
	5''	3	Yes	5050_3_BN
6''	4''	2	No	6040_2
	4''	3	No	6040_3
	4.5''	2	Yes	6045_2_BN
	5''	2	Yes	6050_2_BN

2.1.2. Brushless DC (BLDC) motor. BLDC motor is a synchronous electric motor with electronic commutation system instead of mechanical commutation (brushes in DC motor). Additionally, the electromagnets (armature) remains static while permanent magnets rotate. This type of design allows greater efficiency and less maintenance with respect to brushed (DC) motors. The research evaluates three different types of motors with three different manufacturers for each type.

Table 2. Considered BLDC motors.

Motor	K_v	Propeller diameter	Mass
1306	4000	3-4''	13 g
1806	2700	4-5''	23 g
2205	2300	5-6''	30 g

Motor number specifies the physical dimensions of the rotor. K_v is voltage constant depends on motor's design, and relates voltage, induced emf voltage and rotor angular velocity.

2.1.3. Electronic speed controller (ESC). ESC is integrated inverter/switching power supply which provide three phase bi-directional square voltage to drive the motor. Commutation depends on the rotor position which can be determined by sensored or sensorless methods. Motor speed is affected duty-cycle variation (chopping of supply voltage). With higher PWM frequency, phase voltage increases and as consequence angular velocity of motor increases.

2.1.4. Lithium-polymer (LiPo) battery. Multirotor is high energy consumption UAV. When choosing a battery, it is necessary to consider to several parameters, including relationship between mass and capacity. The main characteristics of the lithium-polymer (LiPo) battery is a very high energy density i.e. the ratio of stored energy per mass. The nominal voltage of individual LiPo battery cell is 3.7V. Fully charged cell has a voltage of 4.2V. Emptying the cells below 3V is not recommended due to the risk of damage and inflammation. Therefore, the possibility of monitoring the battery voltage during the multirotor autonomous flight is extremely important [14]. Research will consider 2S, 3S and 4S battery with high discharge rate as presented in figure 3.

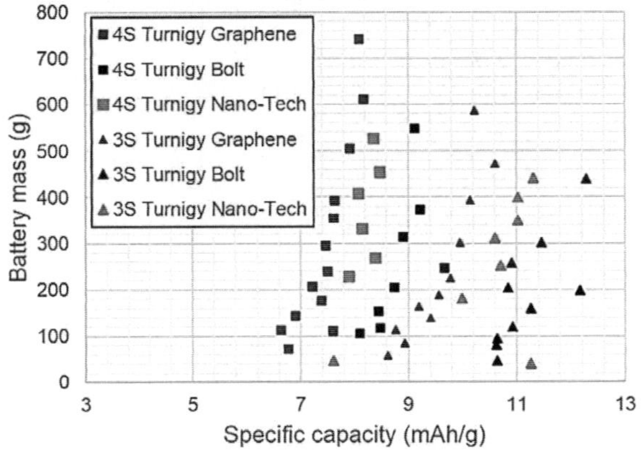

Figure 3. Plot for rechargeable LiPo batteries with high discharge rates [15].

2.2. Multirotor propulsion configuration

The experimental setup allows testing of three various propulsion configurations that could be used for multirotor design. Configurations are defined with rotors arrangement. Single, coaxial and overlapping rotor arrangement can be evaluated.

For single configuration, influence of propeller geometric characteristics on aerodynamic forces will be investigated. For the propeller with the same diameter, influence of various pitch, blade number and so called bull nose design will be examined. Experiment will also compare three BLDC motors of the

IOP Conf. Series: Journal of Physics: Conf. Series **870** (2017) 012003 doi:10.1088/1742-6596/870/1/012003

same geometry and K_V. Measurements will be conducted for 2, 3 and 4 cell batteries. As shown in figure 4, single configuration can be puller or pusher version.

Therefore, analysis of single propulsion yields valuable information regarding influence of specific components combination on overall performance. More efficient combinations of motor and propeller where tested in coaxial configuration (see in figure 5) where rotor flow interaction occurs and have a significant influence to coaxial propulsion efficiency. Therefore, combination of pusher and puller propeller with the same diameter, and various geometric features will be tested.

After testing the coaxial propulsion and propulsion efficiency analysis, so-called overlapping configuration (see in figure 6) will be examined and influence of the overlap area on aerodynamic forces and efficiency where investigated.

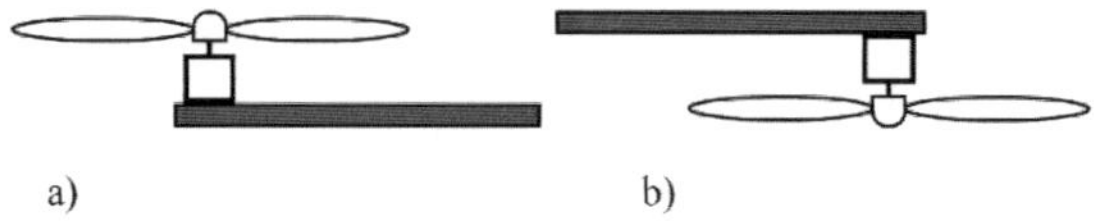

Figure 4. Single rotor arrangement: a) puller, b) pusher.

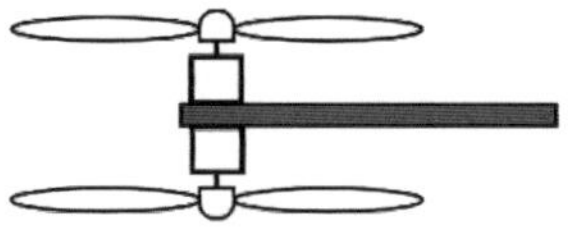

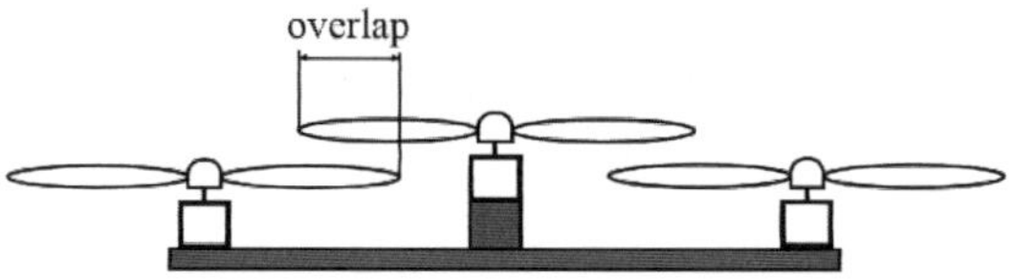

Figure 5. Coaxial rotor arrangement. **Figure 6.** Overlapping rotor arrangement.

3. Experimental setup

The experimental setup for the identification of propulsion physical parameters (see in figure 7) consist from the stand, DC power supply control unit and measurement equipment (as shown in figure 8) which enables data acquisition important for subsequent processing and analysis. Control unit enables simple programming and provides required control signals for identification and further analysis.

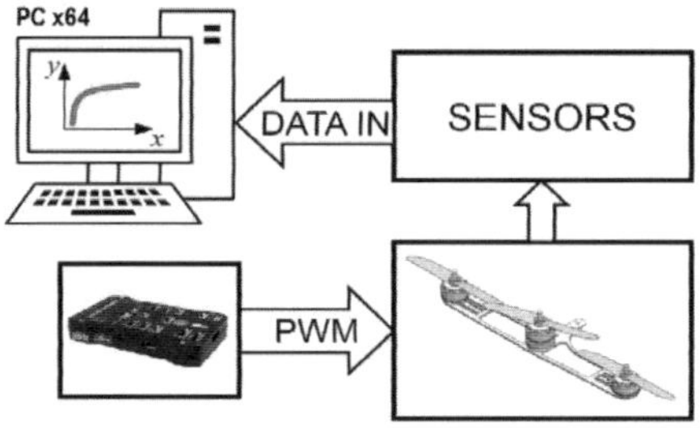

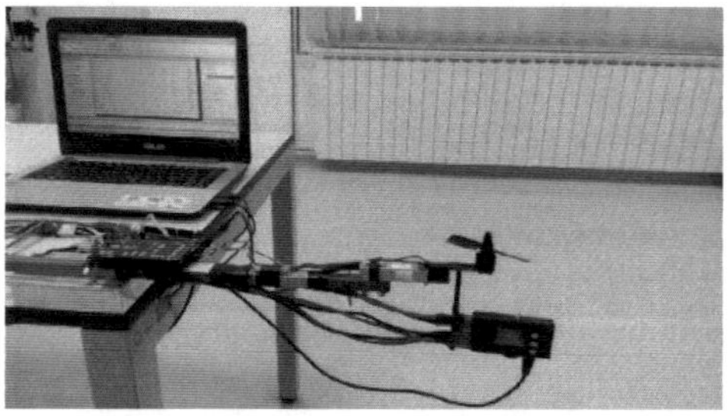

Figure 7. Experimental setup scheme. **Figure 8.** Experimental test bench.

For each propulsion combination, four measurements will be conducted.

3.1. Sensors

To identify the propulsion physical parameters, two sensors where used. Aerodynamic forces where measured with load cell type sensor, while multifunctional logging meter is used to measure angular velocity and power consumption (see in figure 9).

3.1.1. Logging meter. One of the requirements to the design of the experimental setup is simple acquisition and processing of sensor data. Among other things, compact multifunctional and logging meter "PowerLog 6S" is selected because of its simple integration with PC, support for various logging software, 32 bit ARM processor and 12 bit A/D convertor. Also, it provides non-contact optical sensing

IOP Conf. Series: Journal of Physics: Conf. Series **870** (2017) 012003 doi:10.1088/1742-6596/870/1/012003

for propeller revolutions per minute (RPM) measurement with settable blade number and measurement of motor voltage constant [16].

Table 3. PowerLog 6S specifications [16].

Parameter	Value
Input voltage range	4.5 – 60 VDC
Current measurement range	-40 A – +40 A
Cell voltage range	0.05 – 28 VDC
Pulse measurement range	10 μs –999999 μs
Pulse width output range	0–20 ms (0.5 μs step)
RPM measurement range	0–99999 RPM
Voltage resolution / accuracy	0.001V / 0.5%
Current resolution / accuracy	0.01A / 5%
Current loading of test	12 mA
Sample logging time interval	0.25 – 3600 s

3.1.2. Force sensor. Load cell sensor is composed of a strain gauge and an amplifier HX 711 as shown in figure 10. HX711 is a precision 24-bit analog-to-digital converter (ADC), designed for weigh scales and industrial control applications to interface directly with a bridge sensor.

The force sensor where calibrated using precision digital scale (Kern PCB 1000-2) in a range for scheduled measurements. Appropriate control software for measuring and calibration where developed and implemented in Atmel microcontroller.

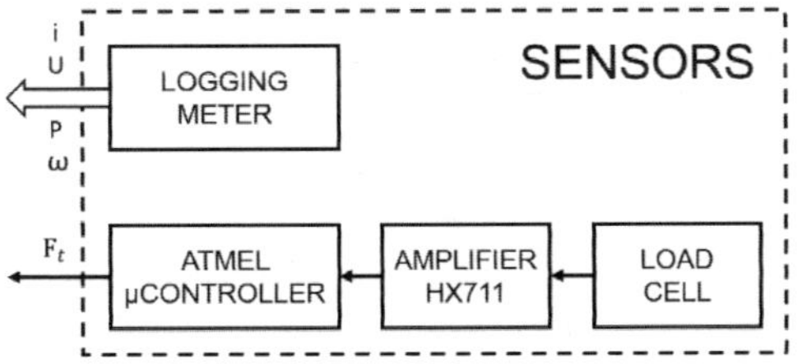

Figure 9. Sensors scheme.

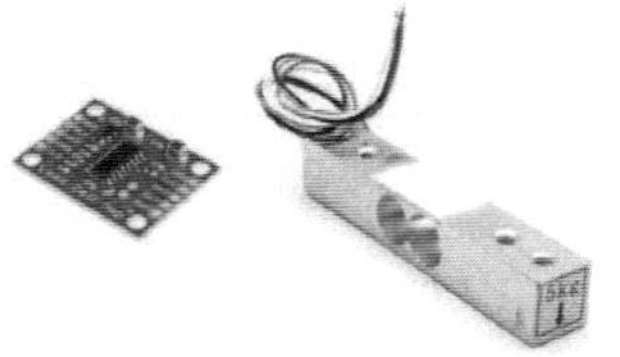

Figure 10. Load cell with amplifier HX711.

3.2. Data Acquisition

Acquisition and fusion of signals (signal from load cell and signals from logging meter) where captured by LogView Studio software, and exported in appropriate format for analysis in MATLAB.

Signal from HX711 analog-to-digital converter where processed in Atmel ATxmega microcontroller. Data is transmitted by USB connection to virtual serial port, in open zero sensor format, to capture data in LogView. Example of a raw vector obtained from the force sensor is shown in figure 11.

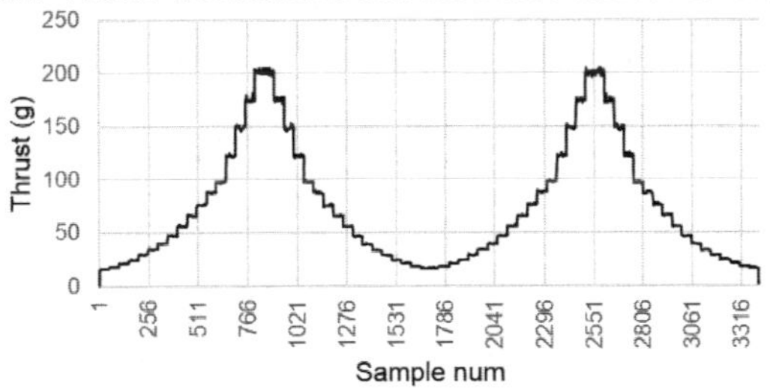

Figure 11. Measured thrust raw vector.

IOP Conf. Series: Journal of Physics: Conf. Series **870** (2017) 012003 doi:10.1088/1742-6596/870/1/012003

3.3. Data Analysis
The data obtained by the sensors where analysed in MATLAB software. Input data represent vectors which are then processed and from which the average values are obtained. The end result are dependences of aerodynamic forces, the angular velocity of the rotor and power consumption in relation to a control signal given from the control unit. From these data, aerodynamic coefficients and propulsion efficiency where obtained which can be used for further design optimization and proper components selection. The results of experimental measurements are presented in section 4.

4. Analysis and Characterization
In order to validate experimental procedure, measurement results for the single configuration propulsion are presented. For this purpose, propeller with 3" diameter and various geometric characteristics in combination with 1306 BLDC motor and 3S battery where used.

4.1. Identified physical parameters
The identified parameters are presented as a function of the signal from the control unit. Parameter values are obtained as the average values of four measurements (see in figures 12-14).

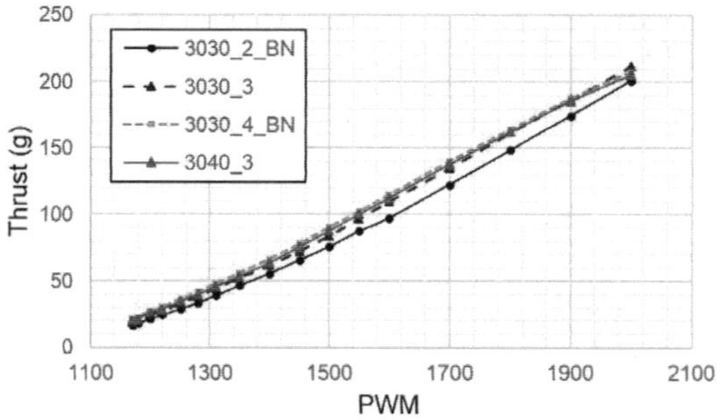

Figure 12. Thrust as a function of the control signal.

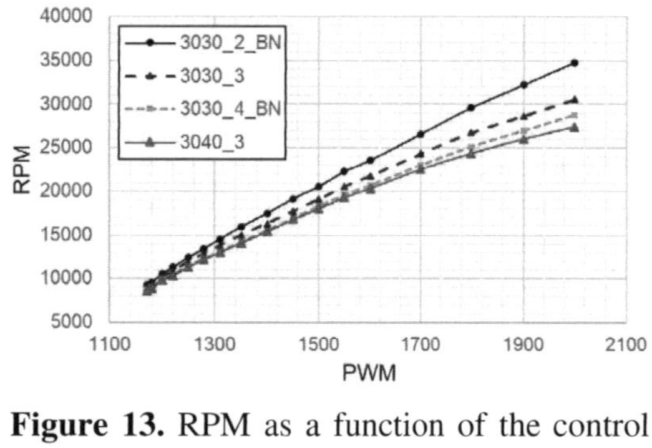

Figure 13. RPM as a function of the control signal.

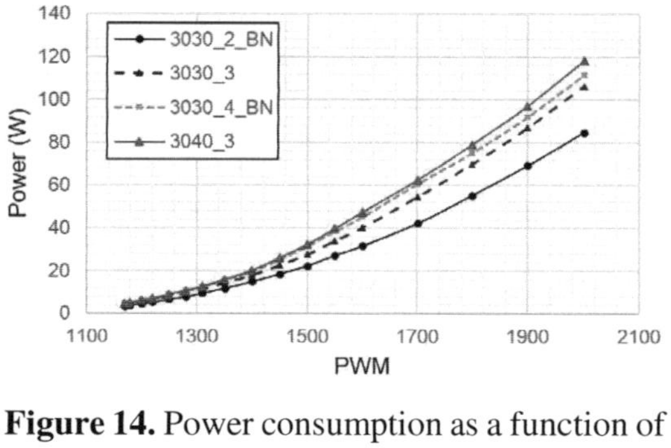

Figure 14. Power consumption as a function of the control signal.

4.2. Aerodynamic force coefficients
For dynamics simulation and control design it is necessary to know thrust and drag coefficients. Using equation (1) and identified thrust and RPM, it is possible to estimate thrust coefficient (see in figure 15 and figure 16). The same procedure applies to the drag coefficient estimation with certain adjustments of experimental setup hardware.

4.3. Propulsion efficiency
The comparison of thrust to consumed power of propulsion unit for different propellers is shown in figure 17. Performance of various propulsion configurations where evaluated based on the efficiency factor which shows how much thrust can be achieved with 1W of electric power (see in figure 18).

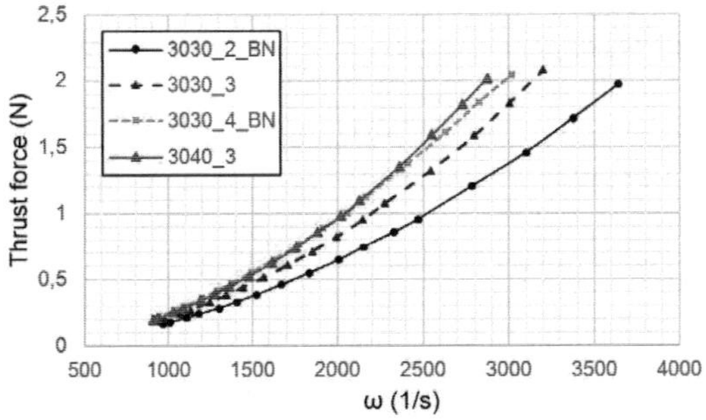

Figure 15. Thrust force as a function of the angular velocity.

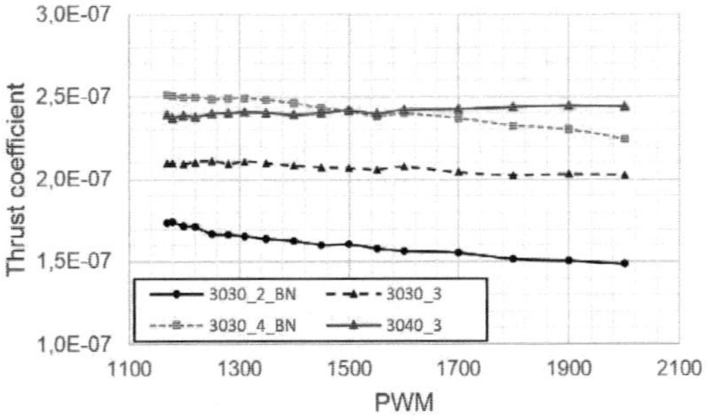

Figure 16. Thrust coefficient.

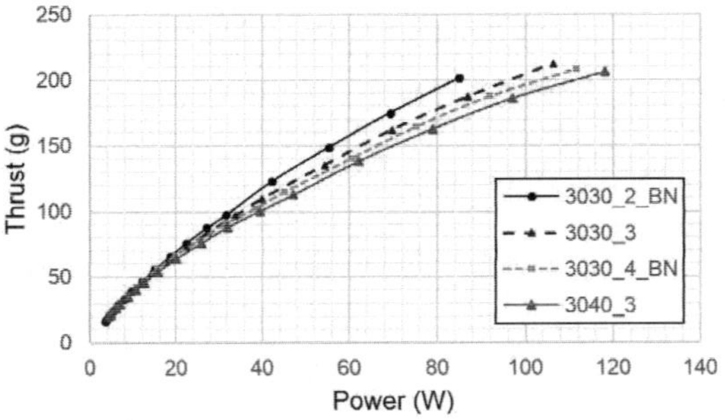

Figure 17. Thrust as a function of the consumed power.

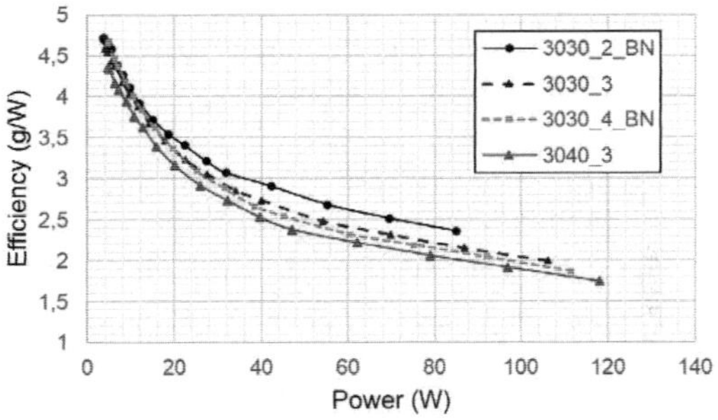

Figure 18. Propulsion efficiency.

5. Conclusion

In this paper, the experimental setup and procedures for identification and analysis of physical parameters for multirotor UAV propulsion is presented. In addition to multirotor design, the greatest impact on the system dynamics and performance have the configuration of propulsion. Therefore, for a successful control design and optimized aircraft design it is necessary to know all the influential physical characteristics. Experimental setup and procedure have been validated in the case of single propulsion unit for combination of 3S battery, 20A ESC, 1306 BLDC motor and propellers with different geometrical characteristics.

Further work will include an analysis of the various propulsion configurations for the proposed components in order to optimize multirotor design and estimation of aerodynamic coefficients and factors.

References

[1] Kumar V and Michael N 2012 Opportunities and challanges with autonomus micro aerial vehicles *Int. J. Rob. Res.* **31**(11) 1279–91

[2] Mizutani S, Okada Y, Salaan C J, Ishii T, Ohno K and Tadokoro S 2015 Proposal and experimental validation of a design strategy for a UAV with a passive rotating spherical shell, *Int. Conf. on Intelligent Robots and Systems (Hamburg)* p 1271–8

[3] Briod A, Kornatowski P, Zufferey J C and Floreano D 2014 A collision-resilient flying robot *J. Field Rob.* **31**(4) 496–509

[4] Bouabdallah S, Murrieri P and Siegeart R 2004 Design and control of an indoor micro quadrotor, *Proc. of the Int. Conf. on Robotics and Automation (New Orleans)*, vol 5 p 4393–8

[5] Mulgaonkor Y, Whitzer M, Morgan B, Kroninger C M, Harrington A M, Kumar V 2014 *Micro- and Nanotechnology Sensors, Systems, and Applications VI*, ed George T, Saif Islam M, Dutta A K, *Proc. of SPIE*, vol 9083

[6] Cheron C, Dennis A, Semerjyan V and Chen Y 2010 A multifunctional HIL testbed for multirotor VTOL UAV actuator, *Int. Conf. on Mechatronics and Embedded Systems and Applications*, p 44-48

[7] Prior S D and Bell J C 2011 Empirical measurements of small unmanned aerial vehicle co-axial rotor systems *J. Sci. Innov.* **1**(1) 1–18

[8] Simoes S M 2015 Optimizing a coaxial propulsion system to a quadcopter, https://fenix.tecnico.ulisboa.pt/downloadFile/563345090412782/Resumo.pdf

[9] Bondrya A, Gardecki S, Gasior P and Giernacki W 2016 Performance of coaxial propulsion in design of multi-rotor UAVs, *Challenges in automation, robotics and measurement techniques (Advances in Intelligent Systems and Computing)*, ed Szewczyk R et al. (Springer International Publishing Switzerland) vol 440 p 523–31

[10] Otsuka H and Nagatani K 2016 Thrust loss saving design of overlapping rotor arrangement on small multirotor unmanned aerial vehicles, *Int. Conf. on Robotics and Automation*, p 3242–8

[11] Theys B, Dimitriadis G, Hendrick P and De Schutter J 2016 Influence of propeller configuration on propulsion system efficiency of multi-rotor unmanned aerial vehicles, *Int. Conf. on Unmanned Aircraft Systems (Arlington)*, p 195–201

[12] Gatti M, Giulietti F and Turci M 2015 Maximum endurance for battery-powdered rotary-wing aircraft *Aero. Sci. Tech.* 45 174–9

[13] Abdilla A, Richards A and Burrow S 2015 Power and endurance modelling of battery-powdered rotorcraft, *Int. Conf. on Intelligent Robots and Systems (Hamburg)*, p 675–80

[14] Brezak H 2016 Robust control of unmanned aerial vehicle with four rotors, Department of Robotics and Production System Automation, Faculty of Mechanical Engineering and Naval Architecture, University of Zagreb

[15] Hobbyking Li-Poly (All brands) 2016 https://hobbyking.com/en_us/batteries/lipoly-all-brands.html

[16] PowerLog 6s Multifunctional Monitor&Logger User's Manual 2016 https://www.progressiverc.com/media/PowerLog%206S%20Manual.pdf

IOP Conf. Series: Journal of Physics: Conf. Series **870** (2017) 012004 doi:10.1088/1742-6596/870/1/012004

Detection of grapes in natural environment using HOG features in low resolution images

Pavel Škrabánek and Filip Majerík

Faculty of Electrical Engineering and Informatics, University of Pardubice, Studentská 95, 532 10 Pardubice, Czech Republic

pavel.skrabanek@upce.cz

Abstract. Detection of grapes in real-life images has importance in various viticulture applications. A grape detector based on an SVM classifier, in combination with a HOG descriptor, has proven to be very efficient in detection of white varieties in high-resolution images. Nevertheless, the high time complexity of such utilization was not suitable for its real-time applications, even when a detector of a simplified structure was used. Thus, we examined possibilities of the simplified version application on images of lower resolutions. For this purpose, we designed a method aimed at search for a detector's setting which gives the best time complexity vs. performance ratio. In order to provide precise evaluation results, we formed new extended datasets. We discovered that even applied on low-resolution images, the simplified detector, with an appropriate setting of all tuneable parameters, was competitive with other state of the art solutions. We concluded that the detector is qualified for real-time detection of grapes in real-life images.

1. Introduction

The latest findings in computer vision were implemented in many agricultural applications and viticulture was no exception. One of the key applications within viticulture was detection of grapes in real scene images. This issue was important in various viticulture applications, such as in autonomous vineyard sprayers, harvesters, or in the process of yield estimation [1-4]. Various image processing, feature extraction and classification algorithms were employed by detection of berries or bunches of grapes in RGB images.

A bunch detector, introduced by Reis at el., utilized colour mapping, morphological dilation, and stem detection [5]. The detector showed correct white wine bunch classification at 90.53 % and red wine at 97.14 %. These results well demonstrated the complexity of white variety detection. Indeed, other solutions aimed at detection of white varieties reached similar results. Berenstein's detector, based on a decision tree algorithm, showed the detection rate of bunches at 90.45 %, and the detection rate of single grapes at 90.10 % [1]. Nuske's solution used a radial symmetry transformation, Gabor filters, and a *k*-nearest neighbours classifier [2]. The detector showed overall precision at 98.00 %, but recall was at only 63.70 %.

A single grape detector, based on the support vector machines (SVMs) classifier, in combination with histograms of oriented gradients (HOG), showed the best performance in white wine varieties detection [6]. Its average accuracy by 10-fold cross-validation (CV) was at 98.23 % for a linear, and at 98.96 % for a radial basis function (RBF) kernel. Moreover, its average precision was at 97.80 % and

IOP Conf. Series: Journal of Physics: Conf. Series **870** (2017) 012004 doi:10.1088/1742-6596/870/1/012004

at 98.49 % for the linear and the RBF kernel, respectively. Its average recall was at 98.68 % and at 99.44 %, respectively. Similar results were achieved by its evaluation on real-life images [7].

Despite the excellent results, the practical utilization of this detector was limited. Indeed, the detector was tested on high-resolution images. Considering the limited computing power of a commonly used hardware in commercial solutions, the detector in its original form could not be used for real-time applications. In order to reduce its time complexity, the structure of the detector was pruned, which resulted in a simplified version of the detector [8]. However, the reduction of the complexity was still insufficient for its use in real-time applications. Here, we address the time complexity vs. performance issue in the context of the image resolution and setting of the detector. We showed that the simplified version of the detector with appropriate setting, guarantees performance comparable with other state of the art solutions.

2. Original works on grape detector

The presented study builds on our previous research published in [6, 8]. Herein, we provide a brief summary of the published detector and datasets.

2.1. Objective of grape detector

The detector was aimed at recognition of grapes of white varieties in object images. RGB object images I (square viewports of source images) of dimensions 40×40 px were considered in [6-8], where px was an abbreviation for a pixel. The resolution of source images was 1936×1288 px, 24 bit.

The detector distinguished between two classes y: 'berry' and 'not berry'. The class 'berry' was called 'positive' and the class 'not berry' was called 'negative'. An object image belonging to the class 'positive' contained a berry of circle shape of diameter ranging between 30 and 40 px. Moreover, the middle of the berry was required to be placed in the middle of the object image with tolerance ±1 px. Object images, which did not satisfy this condition, belonged to the class 'negative'.

2.2. Structure of grape detector

Detection of objects in images usually consists of four successive steps: acquisition of an object image, image pre-processing (IP); extraction of features; and classification. Since the acquisition of object images was not considered in [6, 8], figure 1 shows the complete vision pipeline of the grape detector. Parts of the detector are described in the context of our previous works in further details.

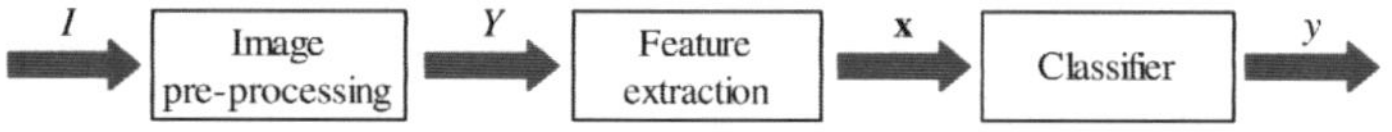

Figure 1. Vision pipeline of grape detectors.

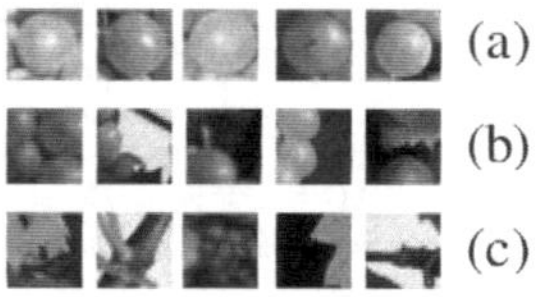

Figure 2. Sample's examples of type: (a) 'positive'; (b) 'negative' in GX, and UX; (c) 'negative' in EX, and UX.

2.2.1. Image pre-processing. Three versions of IPs were introduced. We discovered that a conversion of the input RGB image I to grayscale format Y is the only rational operation to be performed within the IP [8]. The conversion was carried out according to ITU-R recommendation BT.601. This version of the grape detector was named the simplified version.

2.2.2. Feature extraction. The HOG features proved to be convenient for the detection of grapes of white varieties [6]. The feature vector $\mathbf{x}$ was extracted from Y using a HOG descriptor [9]. We used a conservative setting of the detector [6, 8]: linear gradient voting into 9 bins linearly spread over 0 to

180 degrees; cells of size 6×6 px blocks of 2×2 cells; and 1 overlapping cell between adjacent blocks in both directions.

2.2.3. Classifier. Solutions introduced in [6, 8] used SVMs [10] as the classifier. The linear and RBF kernel functions were considered. Regardless of the used kernel function, performance of the classifiers might be influenced by regularization constant C. Performance of the classifier with RBF kernel was predetermined also by a kernel width σ. A grid search algorithm [11] combined with the 10-fold CV was used to find a setting of these parameters giving maximal accuracy [6].

2.3. Training and evaluation

A training set (T-3) of 288 unique 'positive' and 288 unique 'negative' samples was used for the training of the detector [6, 8]. Three evaluation methods were used for the evaluation of the detector [6-8]; however, only the evaluation on test sets is relevant in the context of this paper.

Two types of sets, five sets of each type, were used for the evaluation [6]. The difference between these two types consisted of selection of the 'negative' samples. While 'negative' samples in G (grape type) were composed solely of incomplete berries of diameter between 30 and 40 px, the 'negative' samples in E (environment type) did not capture even the smallest piece of targeted berries.

50 unique 'positive' and 200 unique 'negative' samples were acquired to create a test set. Each test set was extended by artificial 'positive' samples [12] where these samples were created by turning of the images through an angle φ, i.e. the 'positive' samples were turned by $\varphi \in \{0, \pi/2, \pi, 3\pi/2\}$. It means that each test set consisted of 200 'positive' and 200 'negative' samples. All test sets were based on one vineyard row photo which was not used for creating of the training set T-3.

Three performance measures m (acc – accuracy, pr –precision, re – recall) were used in evaluation of the detectors:

$$\mathrm{acc} = \frac{|TP| + |TN|}{|TP| + |FN| + |TN| + |FP|}, \quad \mathrm{pr} = \frac{|TP|}{|TP| + |FP|}, \quad \mathrm{re} = \frac{|TP|}{|TP| + |FN|}, \qquad (1)$$

where $|TP|$ is the number of correctly classified 'positive' samples, $|FN|$ is the number of misclassified 'positive' samples, $|FP|$ is the number of misclassified 'negative' samples, and $|TN|$ is the number of correctly classified 'negative' samples [13].

3. Detector for low–resolution images

In image recognition systems, reduction of the object image resolution brings about a reduction of time-complexity, but often at the expense of the performance. Moreover, the change of the resolution may require significant changes in the image recognition system setting. We addressed all these issues while adapting the grape detector for low-resolution images.

3.1. HOG descriptor properties

Several versions of the HOG descriptor were introduced [9]. The used implementation of the descriptor supports rectangular cells, R-HOG, and spreading of histogram channels over 0 to 180 degrees (unsigned orientation) or 0 to 360 degrees (signed orientation). For the contrast normalization, a L2-hys norm was used. The descriptor has five tuneable parameters: range r (in degrees), number of bins b, size of cells cs (in pixels), number of cells in blocks cb (in cells), and number of overlapping cells between adjacent blocks oc (in cells).

For the used implementation, the range r is a two-valued variable where $r \in \{[0,180],[0,360]\}$ and the number of bins b can be any positive integer. A higher number of bins may ensure encoding finer orientation details, but increasing this value enlarges the feature vector $\mathbf{x}$. Using the signed orientation can help differentiate light-to-dark vs. dark-to-light transitions; however, a higher number of bins b is often needed in order to exploit the potential of the full degrees range.

The remaining three parameters, cs, cb, and oc, are each defined by two values where the first value relates to vertical direction (subscript V) while the second one to horizontal (subscript H). The values can be any positive integer (*oc* can contain also zeros) up to limitations following from image resolutions. Increasing the cell size *cs* or block size *cb* leads to decrease in the length of the feature vector $|\mathbf{x}|$; however, small-scale details may be lost for large cells. And a high block size value reduces the ability to suppress local illumination changes. The number of overlapping cells *oc* influences contrast normalization. To ensure adequate normalization, *oc* should be at least half of *cb*. Large overlap values can capture more information, but they produce a larger feature vector **x**. Thus, the length of the feature vector **x** is

$$|\mathbf{x}| = l_H \cdot l_V \cdot cb_H \cdot cb_V \cdot b, \tag{2}$$

where l is the number of blocks per image in horizontal (l_H) or vertical (l_V) direction. For both directions, the number of blocks can be expressed as

$$l = \left\lfloor \left(|I| \cdot cs^{-1} - cb\right) \cdot \left(cb - oc\right)^{-1} + 1 \right\rfloor \tag{3}$$

where $|I|$ is size of the object image I (in pixels), and $\lfloor \bullet \rfloor$ symbolizes rounding towards minus infinity.

3.2. Search for optimal setting

The simplified version of the grape detector was used as the reliable and efficient platform while developing the solution aimed at detection of grapes in low-resolution images. In order to keep continuity, we used the set T-3 for its training. Considering the relationship between time complexity and performance, we examined possibilities of the detector applications on images resized at 75 %, 50 %, and 25 % of the original size while keeping the aspect ratio 1:1.

The setting of the HOG descriptor is resolution dependent (subsection 3.1); hence, the optimal setting of the HOG descriptor has to be found together with the optimal setting of the classifier for each resolution. Inspired by our previous work [6], we used a one stage grid search algorithm to find the optimal settings. We used two different sets for the training-evaluation process. Within the training phase, we used T-3, while UX (will be introduced in subsection 3.4) was used within the evaluation phase.

Since our goal was to provide a reliable solution with a low time complexity, we searched for a setting giving the best time complexity vs. performance ratio. We formulized the optimization problem as

$$\mathbf{p}^* = \arg\max_{\mathbf{p} \in P} \left(m'(\mathbf{p}) - |\mathbf{x}'(\mathbf{p})| \right) \tag{4}$$

where **p** is a vector describing a setting of all tuneable parameters, P is a set of all considered settings **p**, and **p*** is the optimal setting. The variables m' and $|\mathbf{x}'|$ are normalized performance scores evaluated using the measure m, and normalized length of the feature vector **x**, respectively. We used the feature scaling normalization which can be for the i-th element of a vector **z** written as

$$z_i' = \left(z_i - \min \mathbf{z}\right) \cdot \left(\max \mathbf{z} - \min \mathbf{z}\right)^{-1}. \tag{5}$$

As the measure m, we used acc (1). We considered $\mathbf{p}=(b,cs,cb,r,C)$ for the detector with the linear and $\mathbf{p}=(b,cs,cb,r,C,\sigma)$ for the detector with the RBF kernel. For both directions, H and V, the number of overlapping cells was determined by

$$oc = \lceil 0.5 \cdot cb \rceil, \tag{6}$$

where $\lceil \bullet \rceil$ symbolizes rounding towards plus infinity.

For both kernels and all resolutions, we evaluated the detector for $b \in \{9,10,\ldots,21\}$, $C \in \{1,10,100,1000\}$, and $r \in \{[0,180],[0,360]\}$. For the RBF kernel, the kernel width $\sigma \in \{1,10,100,1000\}$ was considered for all resolutions. The only resolution dependent parameters were the cell size *cs* and the number of cells in blocks *cb*. We considered square cells and blocks of sizes given by limits specified in table 1.

3.3. Evaluation

In order to keep continuity [6-8], we used all three performance measures (1) for the evaluation of the detector with the optimal settings **p***. The evaluation was performed on images of 100 %, 75 %, 50 %,

IOP Conf. Series: Journal of Physics: Conf. Series **870** (2017) 012004 doi:10.1088/1742-6596/870/1/012004

and 25 % of the original size, while keeping the aspect ratio 1:1. For the evaluation, we used new datasets which will be introduced in subsection 3.4.

Table 1. The smallest (min) and the largest (max) sizes of cells *cs* and blocks *cb* used within the grid search for various sizes of object images (first line), where the size of the images is specified as the percentage of the original size 40 × 40 px.

	100 %	75 %	50 %	25 %
cs_{min}	5 × 5	5 × 5	3 × 3	2 × 2
cs_{max}	10 × 10	8 × 8	6 × 6	3 × 3
cb_{min}	2 × 2	2 × 2	2 × 2	2 × 2
cb_{max}	4 × 4	3 × 3	3 × 3	3 × 3

3.4. Extended datasets

The original datasets consisted of 400 (E and G) or 566 samples (T-3). In order to get informative evaluation results, we formed new extended datasets for the search process (UX) and the evaluation (EX and GX). The sets were based on a new collection of thirty unique vineyard row photos (twenty-five for EX and GX and five for UX). The photos were captured under the conditions specified in [6] at 6 different locations.

The sets comprised of labelled RGB object images of size 40 × 40 px which were created from the photos using an editor [14]. 500 unique 'positive' and 2000 unique 'negative' samples were acquired for each set. The sets were extended using the artificial 'positive' samples (see subsection 2.3), i.e. each new dataset consisted of 2000 'positive' and 2000 'negative' samples. The selection of the samples followed the criteria specified in subsection 2.1. The set UX contained both types of negative samples in the ratio 1:1. Examples of the samples are shown in figure 2 for all types of datasets.

The i-th extended test set of type E was denoted as EX-i where $i \in X$ and $X \in \{1,2,\ldots,5\}$. The same labelling principle was used also for the extended test set of type G. Sets EX and GX with the same index i were based on one pool of five photos and they shared a collection of 'positive' samples.

4. Results evaluation

4.1. Search for optimal setting

Solving the optimization problem (4) according to the specifications given in subsection 3.2, we found the optimal settings $\mathbf{p}^*$ which were summarized in table 2. For each optimal setting, we provided the length of the feature vector (table 3, $|\mathbf{x}|^*$), and the performance which was measured using accuracy (table 4, acc^*). For comparison, we presented the smallest ($|\mathbf{x}|_{min}$) and largest ($|\mathbf{x}|_{max}$) lengths of the feature vector $\mathbf{x}$ (table 3), as well as, the best (acc_{max}) and the worst (acc_{min}) performance of the detector according to accuracy (table 4).

Table 2. The optimal settings of the detector for various sizes of object images (first line), and for different kernel functions (first column), where *cs* is cell size, *r* is range, *b* is number of bins, *sb* is block size, *oc* is number of overlapping cells, *C* is regularization constant, and σ is kernel width.

	Linear				**RBF**			
	100 %	75 %	50 %	25 %	100 %	75 %	50 %	25 %
b	11	9	11	13	9	9	9	9
cs	9 × 9	7 × 7	6 × 6	3 × 3	9 × 9	7 × 7	6 × 6	3 × 3
cb	2 × 2	3 × 3	3 × 3	3 × 3	4 × 4	3 × 3	3 × 3	3 × 3
r	[0,180]	[0,180]	[0,360]	[0,360]	[0,180]	[0,180]	[0,360]	[0,360]
C	1	1	1	1	1000	100	10	10
σ	–	–	–	–	100	100	10	10

Table 3. Maximal $|\mathbf{x}|_{max}$, minimal $|\mathbf{x}|_{min}$, and optimal $|\mathbf{x}|^*$ length of the feature vector **x** for various sizes of object images (first line), and for both kernel functions (first column), obtained within the search process performed according to the specification in subsection 3.2.

		100 %	75 %	50 %	25 %
Linear	$\|\mathbf{x}\|_{min}$	144	81	81	81
& RBF	$\|\mathbf{x}\|_{max}$	6804	3024	3024	1701
Linear	$\|\mathbf{x}\|^*$	396	324	99	117
RBF	$\|\mathbf{x}\|^*$	144	324	81	81

Table 4. Best (acc_{max}), worst (acc_{min}) and optimal (acc^*) performance of the detector, measured using accuracy, for various sizes of object images (first line), and for both kernel functions (first column), obtained within the search process performed according to the specification in subsection 3.2.

		100 %	75 %	50 %	25 %
	acc_{min}	0.8318	0.7948	0.8078	0.7460
Linear	acc_{max}	0.9390	0.9138	0.8913	0.8543
	acc^*	0.9288	0.9073	0.8828	0.8543
	acc_{min}	0.5000	0.5000	0.5000	0.5000
RBF	acc_{max}	0.9435	0.9423	0.9085	0.8765
	acc^*	0.9318	0.9245	0.8998	0.8745

4.2. Evaluation

We evaluated performance of the detector, setup according to the specifications summarized in table 2, using accuracy, precession, and recall (1). We summarized the obtained results in table 5 and table 6 for the linear and the RBF kernel, respectively. We provided average values of the measures (bold) for each type of datasets in both tables.

Table 5. Performance of the detector with linear kernel on the extended test sets (first line) for various image sizes (first column), evaluated using accuracy (acc), precision (pr) and recall (re). Average values of the measures are in bold.

		EX-1	EX-2	EX-3	EX-4	EX-5	avg	GX-1	GX-2	GX-3	GX-4	GX-5	avg
	acc	0.9298	0.9218	0.9190	0.9245	0.9168	**0.9224**	0.9365	0.9280	0.9283	0.9295	0.9195	**0.9284**
100%	pr	0.9604	0.9557	0.9505	0.9579	0.9537	**0.9557**	0.9745	0.9688	0.9698	0.9684	0.9595	**0.9682**
	re	0.8965	0.8845	0.8840	0.8880	0.8760	**0.8858**	0.8965	0.8845	0.8840	0.8880	0.8760	**0.8858**
	acc	0.9073	0.8938	0.9043	0.9093	0.8960	**0.9021**	0.9170	0.9040	0.9150	0.9130	0.9020	**0.9102**
75%	pr	0.9483	0.9442	0.9430	0.9585	0.9459	**0.9480**	0.9691	0.9665	0.9658	0.9667	0.9589	**0.9654**
	re	0.8615	0.8370	0.8605	0.8555	0.8400	**0.8509**	0.8615	0.8370	0.8605	0.8555	0.8400	**0.8509**
	acc	0.8878	0.8683	0.8680	0.8820	0.8700	**0.8752**	0.9135	0.8863	0.8865	0.8945	0.8858	**0.8933**
50%	pr	0.9176	0.9216	0.9153	0.9326	0.9209	**0.9216**	0.9715	0.9612	0.9552	0.9598	0.9552	**0.9606**
	re	0.8520	0.8050	0.8110	0.8235	0.8095	**0.8202**	0.8520	0.8050	0.8110	0.8235	0.8095	**0.8202**
	acc	0.8420	0.8285	0.8320	0.8325	0.8348	**0.8340**	0.8825	0.8573	0.8613	0.8590	0.8620	**0.8644**
25%	pr	0.8566	0.8666	0.8597	0.8666	0.8693	**0.8638**	0.9356	0.9261	0.9179	0.9204	0.9249	**0.9250**
	re	0.8215	0.7765	0.7935	0.7860	0.7880	**0.7931**	0.8215	0.7765	0.7935	0.7860	0.7880	**0.7931**

Table 6. Performance of the detector with RBF kernel on the extended test sets (first line) for various image sizes (first column), evaluated using accuracy (acc), precision (pr) and recall (re). Average values of the measures are in bold.

		EX-1	EX-2	EX-3	EX-4	EX-5	avg	GX-1	GX-2	GX-3	GX-4	GX-5	avg
	acc	0.9358	0.9165	0.9228	0.9353	0.9190	**0.9259**	0.9438	0.9243	0.9285	0.9390	0.9235	**0.9318**
100%	pr	0.9589	0.9572	0.9573	0.9658	0.9549	**0.9588**	0.9754	0.9738	0.9693	0.9736	0.9644	**0.9713**
	re	0.9105	0.8720	0.8850	0.9025	0.8795	**0.8899**	0.9105	0.8720	0.8850	0.9025	0.8795	**0.8899**
	acc	0.9273	0.9128	0.9240	0.9263	0.9148	**0.9210**	0.9335	0.9225	0.9325	0.9295	0.9173	**0.9271**
75%	pr	0.9734	0.9640	0.9619	0.9781	0.9689	**0.9693**	0.9871	0.9856	0.9800	0.9853	0.9744	**0.9825**
	re	0.8785	0.8575	0.8830	0.8720	0.8570	**0.8696**	0.8785	0.8575	0.8830	0.8720	0.8570	**0.8696**
	acc	0.9118	0.8910	0.8938	0.9073	0.8898	**0.8987**	0.9225	0.8960	0.9005	0.9088	0.8895	**0.9035**
50%	pr	0.9583	0.9633	0.9555	0.9705	0.9632	**0.9621**	0.9818	0.9748	0.9706	0.9739	0.9626	**0.9727**
	re	0.8610	0.8130	0.8260	0.8400	0.8105	**0.8301**	0.8610	0.8130	0.8260	0.8400	0.8105	**0.8301**
	acc	0.8773	0.8520	0.8610	0.8680	0.8610	**0.8639**	0.8943	0.8660	0.8738	0.8780	0.8650	**0.8754**
25%	pr	0.9289	0.9282	0.9227	0.9391	0.9318	**0.9302**	0.9663	0.9610	0.9511	0.9621	0.9408	**0.9563**
	re	0.8170	0.7630	0.7880	0.7870	0.7790	**0.7868**	0.8170	0.7630	0.7880	0.7870	0.7790	**0.7868**

5. Discussion

We applied the proposed search method for both kernel functions on object images of dimensions 40 × 40 px, 30 × 30 px, 20 × 20 px, and 10 × 10 px. For both kernels and all considered resolutions, the detector setup according to optimal settings $\mathbf{p}^*$ reached near best possible results according to accuracy (table 4) while keeping the length of the feature vector to the smallest possible values (table 3). For all resolutions, the detector with RBF kernel setting had better time complexity vs. performance ratio (based on comparison of $|\mathbf{x}|^*$ and acc*).

Together with the resolution, accuracy of the detector progressively decreased for both kernel functions (tables 4-6). A very similar trend was also observed on precision for the linear kernel, and on recall for the RBF one (tables 5-6). The positive correlation was also observed on recall for both kernels, but the trend was almost linear. The only exception was precision of the detector with the RBF kernel (table 6) where small improvement was observed for object images of sizes 30 × 30 px and 20 × 20 px.

In summary, the reduction of time complexity via the reduction of the resolution will be always at the expense of the performance. From our point of view, the detector with the RBF kernel function, applied on object images of dimensions 20 × 20 px (50 % of the original size) is the best choice. With average accuracy about 90 %, average precision about 97 %, and average recall about 83 %, the detector is still competitive with other state of the art solutions. With the length of the feature vector only 81, such implementation of the detector seems to be suitable for real-time applications. For comparison, the HOG descriptor setup according to original specifications (subsubsection 2.2.2) produces vectors of the length 900.

6. Conclusion

The final performance of an image recognition system is determined by many factors. An appropriate setting of all parameters is one of them. The parameters may be introduced in any of the four stages of the vision pipeline. A large number of parameters have to be considered once the HOG descriptor is used. Setting of such a system is not an easy task which was also the case of the grape detector. In order to face this issue, we proposed the method aimed at search for a setting which gives best time complexity vs. performance ratio of image recognition systems based on the HOG features. The application of the method on the grape detector proved the reliability of the method. We therefore expect a wider utilization of this method while developing image recognition systems of similar structure.

Acknowledgements

The work was supported by the Funds of University of Pardubice, Czech Republic (SGS-2017-027). We would like to offer our special thanks to Víno Sýkora s.r.o. company which enabled us to perform experiments in its vineyards.

References

[1] Berenstein R, Shahar O, Shapiro A and Edan Y 2010 Grape clusters and foliage detection algorithms for autonomous selective vineyard sprayer *Intell. Serv. Robot.* **3** pp 233-43

[2] Nuske S, Achar S, Bates T, Narasimhan S and Singh S 2011 Yield estimation in vineyards by visual grape detection *Proc. 2011 IEEE/RSJ Int. Conf. on Intelligent Robots and Systems* (IEEE) pp 2352-8

[3] Diago M, Correa C, Millán B, Barreiro P, Valero C and Tardaguila S 2012 Grapevine yield and leaf area estimation using supervised classification methodology on RGB images taken under field conditions *Sensors* **12** pp 16988-17006

[4] Liu S and Whitty M 2015 Automatic grape bunch detection in vineyards with an SVM classifier *J. Appl. Log.* **13** pp 643–53

[5] Reis M et al. 2012 Automatic detection of bunches of grapes in natural environment from color images *J. Appl. Log.* **10** pp 285–90

[6] Škrabánek P and Runarsson T 2015 Detection of grapes in natural environment using support vector machine classifier *Proc. of the 21st Int. Conf. on Soft Computing MENDEL 2015* (Brno: Brno University of Technology) pp 143-50

[7] Škrabánek P and Majerík F 2016 Evaluation of performance of grape berry detectors on real-life images *Proc. of the 22nd Int. Conf. on Soft Computing MENDEL 2016* (Brno: Brno University of Technology) pp 217-24

[8] Škrabánek P and Majerík F 2016 Simplified version of white wine grape berries detector based on SVM and HOG features *Artificial Intelligence Perspectives in Intelligent Systems: Proc. of the 5th Computer Science On-line Conf. 2016* Vol 1 (Springer International Publishing) pp 35-45

[9] Dalal N and Triggs B 2005 Histograms of oriented gradients for human detection *IEEE Computer Society Conf. on Computer Vision and Pattern Recognition, 2005* Vol 1 (IEEE) pp 886-93

[10] Burges C 1998 A Tutorial on support vector machines for pattern recognition *Data Min. Knowl. Discov.* **2** pp 121-67

[11] Bergstra J and Bengio J 2012 Random search for hyper-parameter optimization *JMLR* **13** pp 281-305

[12] Lampert C 2008 Kernel methods in computer vision *Found. Trends. Comput. Graph. Vis.* **4** pp 193-285

[13] Sokolova M and Lapalme G 2009 A systematic analysis of performance measures for classification tasks, *Inf. Process. Manage.* **45** pp 427-37

[14] Škrabánek P 2015 Editor for marking and labeling of object images for binary supervised classification in MATLAB environment *Proc. of the 21st Int. Conf. on Soft Computing MENDEL 2015* (Brno: Brno University of Technology) pp 151-58

Impact of corpus domain for sentiment classification: An evaluation study using supervised machine learning techniques

Redouane Karsi[1], Mounia Zaim[1] and Jamila El Alami[1]

[1] Laboratory of System Analysis, Information Processing and Integrated Management, Mohammadia School of Engineers, Mohammed V University, Rabat, Morocco

E-mail: rdkarsi@yahoo.fr, zaim.mounia@yahoo.fr, alamijamila1@gmail.com

Abstract. Thanks to the development of the internet, a large community now has the possibility to communicate and express its opinions and preferences through multiple media such as blogs, forums, social networks and e-commerce sites. Today, it becomes clearer that opinions published on the web are a very valuable source for decision-making, so a rapidly growing field of research called "sentiment analysis" is born to address the problem of automatically determining the polarity (Positive, negative, neutral,...) of textual opinions. People expressing themselves in a particular domain often use specific domain language expressions, thus, building a classifier, which performs well in different domains is a challenging problem. The purpose of this paper is to evaluate the impact of domain for sentiment classification when using machine learning techniques. In our study three popular machine learning techniques: Support Vector Machines (SVM), Naive Bayes and K nearest neighbors(KNN) were applied on datasets collected from different domains. Experimental results show that Support Vector Machines outperforms other classifiers in all domains, since it achieved at least 74.75% accuracy with a standard deviation of 4,08.

1. Introduction

Nowadays, companies and organizations that provide public services increasingly need to extract strategic and decision information from data that has long been produced only through structured information systems, Thus data mining tools have been widely used for data analysis. With the emergence and evolution of social networks, blogs, discussion forums and e-commerce sites, today there is an explosion in the creation of massive data through user-generated web content, these data particularly reflect the preferences and opinions of users in various fields such as politics, health, education, movies, finance, and many research studies have found that the analysis of opinions generated by users is very useful For the prediction of stock price evolution and elections results. Thus, sentiment analysis becomes a very hot field of research [1].

Sentiment analysis is the natural language processing techniques for determining polarity of a text data. The main approaches for sentiment classification are machine learning and lexicon-based approaches. Users express their opinions towards different topics and domains. Thus the challenge is to propose classification methods witch guaranties good performance when applied to different domains, tackling this challenge appears tedious since two different domains use different expressions, for example "honest candidate" expression is frequently used in political domain, however it rarely appears in hotel

domain. Thus, a classifier trained in hotel domain will not be accurate in political domain [2], to overcome this problem many searchers have modelled cross-domain sentiment classification systems which enable training a classifier from one or many domains and applying the trained classifier on a different domain. However, in reality it is not practicable to ensure relatedness between features of the source domain and a large number of target domains, therefore sentiment classification methods are performing well when applied to the same domain as the training data [3].
In this paper supervised machine learning performance was investigated through different domains, choosing to explore machine learning approach is motivated by the fact that searches found that it is more accurate than lexicon based approach when training and testing data are from the same domain [4].
Our contribution is to determine the performance of different machine learning methods applied to different domains, and to explore the effect of varying domain on the accuracy of classifiers by performing experiments using three supervised machine learning: Support Vector machines (SVM), Naive Bayes (NB) and k-nearest neighbors (KNN) applied to distinct datasets covering four different domains: Book, Electronic, Movie and Automobile. Related work on the use of supervised machine techniques for sentiment classification is presented in section 2, our proposed approach is described in section 3, results are commented and discussed in section 4, finally, a conclusion of the paper and an overview of future work are given in section 5.

2. Related Work

Many researchers have investigated supervised machine learning techniques for sentiment classification. From movie reviews dataset, Pang, Lee and Vaithyanathan [5] experimented three machine learning methods (Naive Bayes, maximum entropy and SVM) over n-gram features, they concluded that Naive Bayes shows poor results, whereas SVM algorithm performs good performance. Tripathy, Agrawal and Rath [6] chose to classify movie reviews using Naive Bayes (NB) and Support Vector Machines (SVM), then they compared their performance with results obtained in literature, they observed that SVM yields the best results. Musabah Alkalbani, Mohamed Ghamry and Khadeer Hussain [7] applied five models that are based on five algorithms, the Support Vector Machines algorithm, Naive Bayes algorithm, Naive Bayes (Kernel) algorithm, k-nearest neighbors algorithm, and the decision tree algorithm to classify SaaS (Software as a service) online reviews, they tested all models with unigrams, Term Frequency Inverse Document Frequency (TF-IDF) weighting scheme, using 3-fold, 5- fold and 10-fold cross-validation, the obtained results shows that SVM gives best accuracy. Travel domain was explored by Ye, Zhang and Law [8], their study consist on mining reviews from travel blogs, for this purpose they compared three supervised machine learning algorithms, namely Naive Bayes, SVM and the character based N-gram model for sentiment classification of the reviews on travel blogs for seven popular travel destinations in the US and Europe, the experimental study indicates that SVM and n-gram achieve better performance than the Naive Bayes Method, in addition, When training data sets had 500 or more reviews, all three approaches could reach accuracies of more than 80%. Shi and Li [9] conducted their study on hotel domain, they experimented a supervised machine learning approach using unigram features with two types of weighting scheme (frequency and TF-IDF) to realize opinion classification of documents, they found that using TF-IDF provides more accurate results. Tripathy, Agrawal and Rath [10] Classified movie reviews with four machine learning algorithms: Naive Bayes (NB), Maximum Entropy (ME), Stochastic Gradient Descent (SGD), and Support Vector Machine (SVM), they found from empirical tests that as the value of "n" in n-gram increases the classification accuracy decreases.

3. Methodology

The steps of our classification approach are shown in figure 1. In our study, three machine learning algorithms were applied, namely Support Vector Machines (SVM), Naive Bayes and K-nearest neighbors (KNN) for sentiment classification of reviews from datasets in different domains.

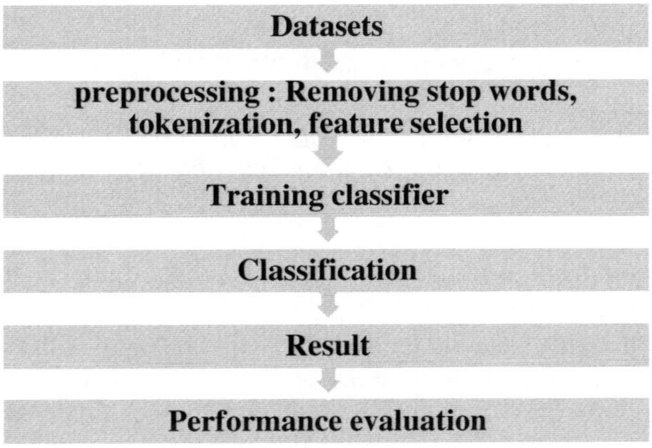

Figure 1. Sentiment classification steps with supervised machine learning.

3.1. Dataset Collection

For our study, four publicly available datasets for research purposes were used, these datasets are described as follows:

Multi-Domain Sentiment Dataset: it is a dataset constructed by Blitzer, Dredze and Pereira [11], they collected Amazon product reviews representing different domains, in our experiment, two domains: Book and Electronic were selected, each domain was built over a collection of 2000 reviews (1000 positive, 1000 negative).

Epinions dataset is built through automobile reviews downloaded from epinions.com web site, the dataset contains 1382 posts with 691 positive and 691 negative reviews.

Polarity dataset: this dataset was created by Pang and Lee [12], it contains 1000 positive and 1000 negative movie reviews all written before 2002.

3.2. Supervised machine learning techniques

In this paper, supervised learning techniques towards different domains were applied by selecting commonly used algorithms for addressing classification problems based on supervised learning methods, in our case, Support vector machines (SVM), Naive Bayes, k-nearest neighbors (KNN) are experimented.

- Support vector machines (SVM): Support Vector Machines (SVM) are new discriminating techniques in statistical learning theory. They have been proposed in 1995 by V. Vapnik in his book "The Nature of Statistical Learning». They allowed tackling several different problems such as regression, Classification… SVM consists in projecting the data of the input space (belonging to two different classes) non-linearly separable in a space of a larger dimension called a feature space so that the data becomes linearly separable. In this space, an optimal hyperplane separating classes such that the two sides of the hyperplane separate the vectors belonging to the different classes and the smallest distance between the vectors and the hyperplane, (the margin) is Maximum [13].

- Naive Bayes: Naive Bayes classifiers are linear and based on Bayes' theorem [10], and the naive term refers to the assumption that words in a dataset are independent. Given an document represented by a dimensional vector $D_i = (X_1,X_2,\ldots,X_n)$, where X_k is the term k of the document D_i, and given a set of m classes,$C_1,C_2,\ldots,C_m$. Using Bayes theorem, the naive Bayesian classifier calculates the posterior probability of each class conditioned on D_i. Di is assigned the class label of the class with the maximum posterior probability conditioned on D_i. Therefore, the most likely class for a document D_i is computed as

$$C^* = argmax_j\, p(Cj)p({}^{Di}/_{Cj}) \quad (1)$$

- $P(C_j)$ is the probability of class C_j.
- $P(D_i/C_j)$ is the probability of document D_i conditioned on C_j.

- K-nearest neighbors (KNN): the idea of KNN algorithm is that the class of an observation X is determined according to a majority vote of the k nearest neighbors of the observation [7].

3.3. Proposed approach

Before applying machine learning algorithms, datasets collected for our experiment are preprocessed by following several steps as schematized in figure 1.

- Stop words removal: Unnecessary and insignificant words are removed from datasets.

- Stemming: each word is transformed into its stem.

- Feature selection: Only unigrams from text was extracted, each selected feature was assigned a weight, in this study a popular weighting scheme, TF-IDF was used, where TF is the number of times a term occurs within a document and TF-IDF witch measure the importance of a feature in the document and the whole corpus is calculated according to the following formula:

$$\mathrm{TFIDF} = \mathrm{TF} * \log(\frac{\mathrm{N}}{\mathrm{df}}) \qquad (2)$$

- N is the total number of documents.
- df is the number of documents in which the term appears.

-Training classifiers: 3-fold, 5- fold and 10-fold cross-validations in training and test processes [7] were used.

- Performance evaluation: To evaluate the performance of the algorithms, evaluation metrics (precision, recall and accuracy) were computed.

$$Precision = \frac{TP}{TP + FP} \qquad (3)$$

$$Recall = \frac{TP}{TP + FN} \qquad (4)$$

$$Accuracy = \frac{TP + TN}{TP + FP + TN + FN} \qquad (5)$$

- TP (True Positive): correctly classified as positive.
- FP (False Positive): incorrectly classified as positive.
- TN (True Negative): correctly classified as negative.
- FN (False Negative): incorrectly classified as negative.

4. Experimental Results

To evaluate the performance of our selected supervised machine learning techniques, multiple performance indicators were computed, namely: Accuracy, recall and precision, for training and testing data, 3, 5, 10 fold cross validation models were used as shown in tables 2 to 5.

After calculating the average accuracy of each algorithm on all domains, SVM shows the best accuracy with an average of 78.92%, followed by NB with 75.95% accuracy, and then KNN achieves a low average of 58.07%.

To evaluate the impact of domain on classification accuracy, the results of our experiment revealed that for SVM the accuracy varies between 74.75% in electronic domain and 85.25% in the movie domain, while NB shows a Minimum accuracy in the book domain with 72.3% and a maximum accuracy of 81.4% in the movie domain, 5NN achieves classification accuracy ranging from 53.5% in the book domain to 64.32% in the automobile domain, 20NN gives accuracies varying between 51.45% in the book domain and 67.22% in the automobile domain, finally 45NN offers an accuracy of 50.3% in the electronic domain and 67.87% in the Automobile domain.

Table 1. Standard deviation of accuracy for applied machine learning algorithms.

Algorithm	Standard deviation
SVM	4.28
NB	4.04
5NN	5.15
20NN	6.74
45NN	8.42

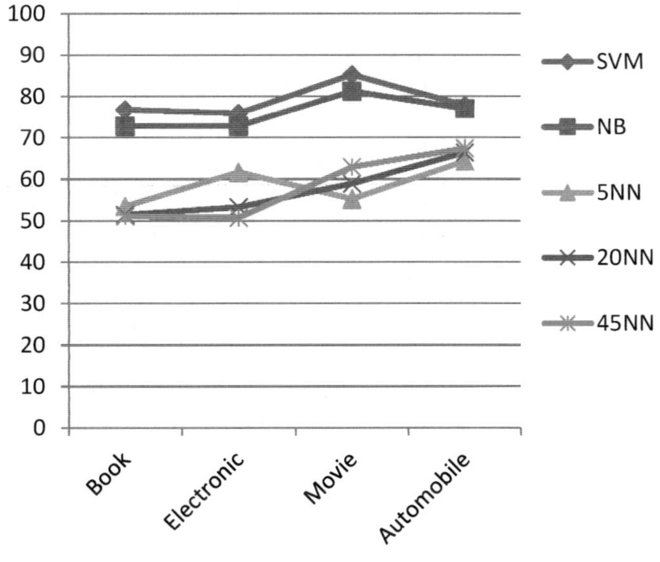

Figure 2. Classification accuracy variation within different domains.

With values of standard deviation less than 5 as shown in table 1, NB and SVM appear to be the most suitable classifiers performing almost similar performances over several domains. Combining low precision and high dispersion, KNN is not good to be used as classifier in different domains.
According to the results of our experiment as shown in figure 2, SVM and NB are more efficient when applied to movie domain, otherwise, KNN behaves well in the automobile domain.
The results also show that k-fold cross validation model affect randomly the classification performance.

Table 2. Performance of different supervised machine learning algorithms on Book domain.

	Fold-3			Fold-5			Fold-10		
	Accuracy	Recall	Precision	Accuracy	Recall	Precision	Accuracy	Recall	Precision
SVM	75.05	75.10	75.90	77.20	77.20	78.00	76.75	76.80	77.40
NB	72.30	72.30	72.30	73.40	73.40	73.40	72.80	72.80	72.80
5NN	54.55	54.60	61.20	53.75	53.80	60.20	53.50	53.50	59.40
20NN	51.85	51.90	58.10	51.70	51.70	63.30	51.45	51,50	62.70
45NN	53.15	53.20	61.90	51.85	51.90	70.20	51.10	51.10	68.60

Table 3. Performance of different supervised machine learning algorithms on Electronic domain.

	Fold 3			Fold 5			Fold 10		
	Accuracy	Recall	Precision	Accuracy	Recall	Precision	Accuracy	Recall	Precision
SVM	74.75	74.80	77.60	77.05	77.10	79.10	75.90	75.90	76.00
NB	75.15	75.20	75.30	75.75	75.80	75.80	72.80	72.80	72.80
5NN	59.00	59.00	59.60	61.10	61.10	61.60	61.70	61.70	62.10
20NN	52.90	52.90	64.40	54.10	54.10	65.50	53.25	53.30	63.90
45NN	50.30	50.30	65.10	50.55	50.60	75.10	50.75	50.80	72.20

Table 4. Performance of different supervised machine learning algorithms on Movie domain.

	Fold 3			Fold 5			Fold 10		
	Accuracy	Recall	Precision	Accuracy	Recall	Precision	Accuracy	Recall	Precision
SVM	83.85	83.90	83.90	84.20	84.20	84.20	85.25	85.30	85.30
NB	81.25	81.30	81.50	81.40	81.40	81.70	81.25	81.30	81.50
5NN	57.15	57.20	57.20	54.45	54.50	54.60	55.20	55.20	55.50
20NN	57.65	57.70	57.80	58.40	58.40	58.50	58.95	59.00	59.30
45NN	61.40	61.40	62.00	60.75	60.80	60.90	62.90	62.90	62.90

Table 5. Performance of different supervised machine learning algorithms on Automobile domain.

	Fold 3			Fold 5			Fold 10		
	Accuracy	Recall	Precision	Accuracy	Recall	Precision	Accuracy	Recall	Precision
SVM	75.54	75.50	76.80	76.91	76.90	77.80	77.78	77.80	78.30
NB	76.84	76.80	76.90	76.41	76.40	76.40	76.98	77.00	77.10
5NN	64.25	64.30	67.00	63.02	63.00	65.50	64.32	64.30	66.70
20NN	67.22	67.20	71.10	65.48	65.50	69.10	66.42	66.40	70.60
45NN	67.80	67.80	72.10	67.87	67.90	73.10	67.43	67.40	72.90

5. Conclusion

The aim of this paper was to investigate the impact of domain on classification performance by proceeding with a machine learning-based approach tested on publicly available datasets covering Book, Electronic, Movie and Automobile domains, According to the results of our experimental study, it was found that Support Vector Machines (SVM) is the most appropriate classifier to be used on different domains.

For further work, it will be examined how the size of the dataset affects classification performance, for that, several datasets of different sizes collected from different domains will be tested.

References

[1] W. Fangzhao, H. Yongfeng and Y. Zhigang, "Domain-Specific Sentiment Classification via Fusing Sentiment," *Information Fusion,* 2016.

[2] B. Heredia, T. M. Khoshgoftaar, J. Prusa and M. Crawford, "Cross-Domain Sentiment Analysis: An Empirical Investigation," in *IEEE 17th International Conference on Information Reuse and Integration*, 2016.

[3] J. Carroll, D. Bollegala and D. Weir, "Cross-Domain Sentiment Classification using a Sentiment Sensitive Thesaurus," *IEEE Transactions on Knowledge and Data Engineering,* vol. **25**, pp. 1719-1731, 2013.

[4] D. Shuyuan, P. S. Atish and Z. Huimin, "Adapting sentiment lexicons to domain-specific social media texts," *Decision Support Systems,* vol. **94**, p. 65–76, 2017.

[5] B. Pang, L. Lee and S. Vaithyanathan, "Thumbs up? Sentiment Classification using Machine Learning Techniques," 2002.

[6] A. Tripathy, A. Agrawal and S. K. Rath, "Classification of Sentimental Reviews Using Machine Learning," *Procedia Computer Science,* vol. **57**, pp. 821-829, 2015.

[7] A. Musabah Alkalbani, A. Mohamed Ghamry and F. Khadeer Hussain, "Predicting the sentiment of SaaS online reviews using supervised machine learning techniques," in *International Joint Conference on Neural Networks (IJCNN)*, 2016.

[8] Q. Ye, Z. Zhang and R. Law, "Sentiment classification of online reviews to travel destinations by supervised machine learning approaches," *Expert Systems with Applications,* vol. **36**, p. 6527–6535.

[9] H.-X. Shi and X.-J. Li, "A sentiment analysis model for hotel reviews based on supervised learning," in *2011 International Conference on Machine Learning and Cybernetics*, 2011.

[10] A. Tripathy, A. Agrawal and S. K. Rath, "Classification of sentiment reviews using n-gram machine learning approach," *Expert Systems with Applications,* vol. **57**, p. 117–126, 2016.

[11] J. Blitzer, M. Dredze and F. Pereira, "Biographies, Bollywood, Boom-boxes and Blenders: Domain Adaptation for Sentiment Classification," in *Proceedings of ACL-07, the 45th Annual Meeting of the Association of Computational Linguistics*, 2007.

[12] B. Pang and L. Lee, "A Sentimental Education: Sentiment Analysis Using Subjectivity Summarization Based on Minimum Cuts," *Proceedings of the 42nd ACL,* pp. 271-278, 2004.

[13] T. Mullen and N. Collier, "Sentiment analysis using support vector machines with diverse information sources," in *Conference on Empirical Methods in Natural Language* , 2004.

Dynamic mapping of EDDL device descriptions to OPC UA

Kofi Atta Nsiah, Manuel Schappacher, and Axel Sikora

Institute of Reliable Embedded Systems and Communication Electronics (ivESK), Offenburg University of Applied Sciences, 77652 Offenburg, Germany

kofi.nsiah@hs-offenburg.de

Abstract. OPC UA (Open Platform Communications Unified Architecture) is already a well-known concept used widely in the automation industry. In the area of factory automation, OPC UA models the underlying field devices such as sensors and actuators in an OPC UA server to allow connecting OPC UA clients to access device-specific information via a standardized information model. One of the requirements of the OPC UA server to represent field device data using its information model is to have advanced knowledge about the properties of the field devices in the form of device descriptions. The international standard IEC 61804 specifies EDDL (Electronic Device Description Language) as a generic language for describing the properties of field devices. In this paper, the authors describe a possibility to dynamically map and integrate field device descriptions based on EDDL into OPCUA.

1. Introduction

In today's automation industry, the increasing number and types of field devices from different manufacturers require integration into industrial control systems. As a result, the provision of a standardized access to these field devices allows fairly easy device management such as configuration, diagnosis and maintenance. So far, major device description and integration technologies are EDDL (Electronic Device Description Language) [1-3], FDT (Field Device Tool) [1, 2, 4] and FDI (Field Device Integration) [1, 2, 5]. These device integration technologies provide different mechanisms to solve the provision of standardized access to field device problem.

In OPC UA terms, this standardized access is provided by the information model of the server-side of OPC UA's client server architecture. An OPC UA server exposes its data via its information model. In order to represent the data of field devices, the OPC UA server has to have advanced knowledge about the field devices in the form of device descriptions. EDDL, FDT/DTM and FDI technologies define device descriptions and their integration into OPC UA. The device descriptions can either be used to create new information models before the OPC UA server is started or can be used to alter the information model of the already running OPC UA server. In this paper, the authors investigate the dynamic integration of device descriptions by altering the information model of the OPC UA server at run-time.

This paper is organized as follows: Chapter 2 provides a brief overview of the OPC UA technology in terms of its system architecture, information model and comparisons of the existing open source implementations of the technology. Chapter 3 presents the EDDL technology and provides a brief overview of other competing solutions. Finally, chapter 4 provides an insight into the implementation of the dynamic mapping of device descriptions based on the EDDL technology to OPC UA by the authors.

2. OPC UA

OPC UA, the successor to the legacy OPC technology, was developed to address the inherent technical defects associated with the classic OPC technology. Compared to the OPC technology, OPC UA provides a consistent and integrated address space that allows a single OPC UA server to integrate data, historical data, alarms and events [6]. Additionally, OPC UA defines a service model that provides a set of services that allows the data exposed in the address space of the OPC UA server to be accessed by connected OPC UA clients. The contents of the address space are described by well-defined information models.

2.1 OPC UA system architecture

The OPC UA system architecture models OPC UA clients and servers as interacting partners. The OPC UA server provides data by exposing its information model while the OPC UA client consumes the data provided by the information model of the OPC UA server. OPC UA services define the APIs (Application Programming Interfaces) that allow the OPC UA clients to interact with the OPC UA servers. Generally, a typical OPC UA client or server consists of three software layers briefly explained as follows:

- OPC UA Communication stack – implements the different data encoding mechanisms, message security and transport protocols defined by the OPC UA technology.
- OPC UA client or server SDK – implements common OPC UA functionalities including OPC UA services. The SDK is not a mandatory feature, however if implemented, reduces the development efforts of the client or server application.
- Client or Server application – uses the OPC UA communication stack and SDK to send and receive OPC UA messages. Specific functionalities can be implemented in this layer.

2.2 OPC UA information model

The OPC UA information model is based on the OPC UA Meta model [7]. The OPC UA information model provides not only the pure data but also exposes the semantic of the provided data. OPC UA information modelling is always done on the server-side however, they can be accessed and modified by connected OPC UA clients.

The data provided by the information model of OPC UA servers are represented as a set of nodes described by attributes and interconnected by references [8]. This data exposed by the OPC UA server is based on the eight base nodes and their attributes as shown in Figure 1.

Each OPC UA node has a unique purpose. For example, the Variable node is used to model the value of a field device while the DataType node models simple and structured data types of the Variable node's value. Depending on the purpose of an OPC UA node, they can have different sets of attributes. However, there are some attributes that are common to all OPC UA nodes.

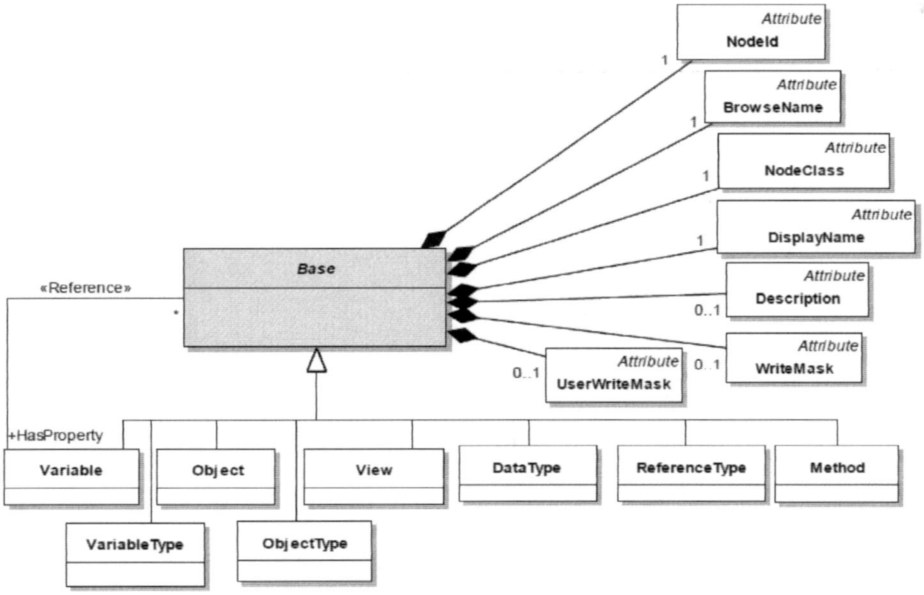

Figure 1. OPC UA Meta Model [7]

3. Device Descriptions

3.1. EDDL

EDDL (Electronic Device Description Language) is a structured text-based device description technology for describing the information that is accessible in digital devices. EDDL is an open technology with international standard status and can be applied to several field devices. The format of the described information is similar to an XML, HTML web page or SGML document. EDDL describes these digital devices in terms of device parameters and their dependencies, device functions, visual representations as well as interactions with control systems [9]. EDD (Electronic Device Description) is the resulting file deployed by device manufacturers that use the EDDL technology to describe their devices. A sample EDD describing the parameters contained in a NIKI current and temperature sensor is as shown in Figure 2.

```
/**
 * Sample EDD showing a sample description of
 * parameters  contained in a current or temperature
 * sensor.
 */

MANUFACTURER       66,
DEVICE_TYPE        0x070E,
DEVICE_REVISION    1,
DD_REVISION        1

/**
 * Describes the NIKI Current Sensor.
 */
VARIABLE        NIKI_Current_Sensor
{
    LABEL        NIKI_Current_Sensor;
    HELP         measures_actual_current;
    CLASS        CONTAINED & DYNAMIC;
    TYPE         FLOAT;
    HANDLING     READ;
}

/**
 * Describes the NIKI Temperature sensor.
 */
VARIABLE        NIKI_Temperature_Sensor
{
    LABEL        NIKI_Temperature_Sensor;
    HELP         measures_actual_temperature;
    CLASS        CONTAINED;
    TYPE         FLOAT;
    {
        DEFAULT_VALUE  0.0;
    }
    HANDLING  READ & WRITE;
}
```

Figure 2. EDD for NIKI Current and Temperature Sensor.

3.2. EDDL as a Data Model

The general motivation of this sub-chapter is to describe the EDDL as a data model. The EDDL technology provides a set of scalable language elements that are used in describing the field devices. EDDL language elements consist of identification elements, basic construction elements and special elements [9]. The identification elements (e.g. MANUFACTURER, DEVICE_TYPE) specify identification information that uniquely identifies the device description of a specific device type from a device manufacturer.

The basic construction elements support device descriptions in terms of device properties and related device functionalities. These elements define a set of associated attributes that refine their definition. Finally, special elements are EDDL extensions that support additional features. [9]. Table 1 presents some commonly used EDDL basic construction elements with their respective descriptions.

IOP Conf. Series: Journal of Physics: Conf. Series **870** (2017) 012006 doi:10.1088/1742-6596/870/1/012006

Table 1. EDDL language elements and their descriptions.

EDDL element	Description
VARIABLE	Describes parameters contained in a device.
METHOD	Defines executable subroutines.
COMMAND	Describes the structure and the addressing of variables in the device.
BLOCK	Describes a field device that is organized in logical blocks.
MENU	Organizes EDDL elements into a hierarchical structure.
LIKE	Used to create a new instance of an existing item (e.g. VARIABLE)
GRAPH	Used to display data from a device.

EDDL technology consists of EDD file and a host PC that hosts an EDDL interpreter. EDD file contains the descriptions of the devices as supplied by the device manufacturers. EDDs supplied in the form of ASCII text files are imported into the host PC, interpreted by the EDDL interpreter and executed.

3.3. EDDL vs FDT/DTM

EDDL and FDT/DTM (Field Device Tool/ Device Type Manager) technologies are two most widely used device description technologies. EDDL technology comprises two components – EDD file, which contains the device descriptions and a host application that reads and executes the EDD file. On the other hand, the FDT/DTM technology consist of two software components – DTM, a software component that contains device descriptions or descriptions of the communication component and FDT Frame Application that provides a common runtime environment for all DTMs [4]. Compared to the FDT/DTM technology, EDDL technology is operating system independent and portable. Recently, the FDI (Field Device Integration) technology has been developed to combine the advantages of both competing solutions in a single solution. The core component of the FDI technology is the FDI Package which contains the device descriptions and other optional components. The FDI device descriptions are based on the EDDL technology [5].

4. Implementation

4.1. Introduction

This section presents the OPC UA open source solutions [10] that were compared by the authors to finally select one for our implementation, presents the EDDL parser implementation and the EDDL to OPC UA mapping. This section also describes shortly the test setup to test whether the authors' approach of dynamically mapping devices based on the EDDL technology to OPC UA is feasible.

4.2 OPC UA open source solutions

Open source solutions of the OPC UA technology developed by the OPC Foundation, other companies, research institutes and academia as well as commercial solutions of the technology are available. In this paper, the authors compare the existing twelve open source implementations [10] based on some selected parameters. These parameters include the open source solution's support for the different OPC UA communication mappings, security profiles, transport profiles, OPC UA services, license types, community updates, programming languages used, operating system and hardware targets as well as tools required to enhance usability of the solutions. A snapshot of the parameters supported by three of the open source solutions, UA.Net [11], open62541 [12], AsNeG [13-14] are as shown in Figure 3.

Comparing the selected parameters of all the open source solutions, the AsNeG (Automation Service Next Generation) solution was preferred by the authors and selected as our reference OPC UA server implementation. AsNeG is open, free and is developed in C++.

IOP Conf. Series: Journal of Physics: Conf. Series **870** (2017) 012006 doi:10.1088/1742-6596/870/1/012006

Parameter		UA-.NET	open62541	AsNeG
Data encoding	OPC UA Binary	✓	✓	✓
	OPC UA XML	✓	×	×
Message security protocols	UA Secure Conversation	✓	✓	✓
Transport protocols	UA-TCP	✓	✓	✓
	HTTPS	✓	×	×
Security Profiles	None	✓	✓	✓
	Basic128Rsa15	✓	×	✓
	Basic256	✓	×	✓
	Basic256Sha256	✓	×	✓
Transport Profiles	UA-TCP UA-SC UA Binary	✓	✓	✓
	HTTPS UA Binary	✓	×	×
	HTTPS UA XML	✓	×	×
Discovery Service Set	FindServers	✓	✓	✓
	GetEndpoints	✓	✓	✓
	RegisterServer	✓	×	✓
SecureChannel Service Set	OpenSecureChannel	✓	✓	✓
	CloseSecureChannel	✓	✓	✓
Session Service Set	CreateSession	✓	✓	✓
	ActivateSession	✓	✓	✓
	CloseSession	✓	✓	✓
	Cancel	✓	×	✓

Figure 3. OPC UA open source solutions

4.3. EDDL Parser

EDD files can be distributed by device manufacturers either in text or binary format depending on what the host system requires. Binary EDD file is normally tokenized to a compressed binary to prevent tampering of the EDD file. These compressed files are relatively smaller, uploaded by the host system and decoded by DD services. On the other hand, text-based EDD files are interpreted by EDDI (Electronic Device Description Interpreter) in the host system where these EDD files are uploaded. In this paper, only text-based EDD files are considered.

The main motivation for implementing the EDDL parser is to extract the EDDL language elements and map them to OPC UA nodes. Several tools and libraries such as GNU Bison – YACC compatible parser generator, Boost Spirit library and Flex are possible solutions to implement the EDDL parser. The EDDL parser implemented by the authors has the following characteristics:

- Based on the Boost Spirit library
- Parses all EDDL language elements (e.g. language elements for identifying a specific EDD, describing data and communication components of field devices).
- Skips all whitespaces, pre-processor directives and all comments in the EDD file
- Parser can be extended to parse additional EDDL language elements that would be required in future.

4.4. EDDL to OPC UA mapping

The FDI technical specification part 5 defines the mapping of EDDL and other FDI package information to the FDI information model and its underlying OPC UA information models [15]. Based on these defined mappings, the authors have designed APIs to automatically map the parsed EDDL language elements or constructs to their corresponding OPC UA nodes or attributes. Even though these mappings are well-defined, not all EDDL language elements have their corresponding mappings. For example, while the EDDL VARIABLE language element maps to a Variable node in OPC UA, the EDDL COMMAND language element has no mapping. The reason is based on the purpose of the EDDL language element. OPC UA clients access Variable nodes in the OPC UA server to read or write sensor values but do not need to access the COMMAND element. As explained earlier, the EDDL COMMAND language element supports the mapping of the EDDL file communication elements to the underlying communication system.

The simple logic flow of how the EDDL to OPC UA mapping is achieved is as shown in Figure 4.

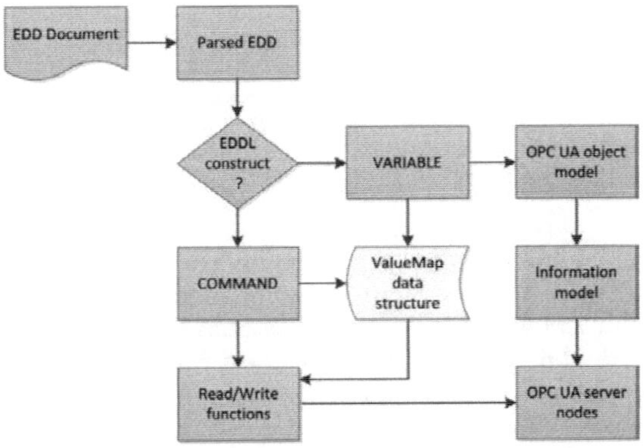

Figure 4. Flowchart showing EDDL to OPC UA mapping

In Figure 6, an EDDL file (EDD Document) with the VARIABLE and COMMAND EDDL language elements is parsed by our own EDDL parser. According to the FDI specification [15], the VARIABLE language element is mapped to create corresponding OPC UA variable nodes in the OPC UA server. The COMMAND language element with no defined mapping however allows the OPC UA server to have knowledge about how data can be read or written to the actual field devices or sensors. In between these operations, a complex data structure is implemented to store all relevant information.

4.5 Test setup

The objective of our test setup is to provide a platform to test the dynamic integration of device descriptions based on the EDDL technology into OPC UA. The test setup consists of an embedded platform, which hosts the server-side of the OPC UA implementation and a Windows-based PC running the client-side of the OPC UA implementation. In our set up, the OPC UA server open source implementation of the AsNeG solution is selected, fine-tuned and run on the embedded platform. On the other hand, UaExpert developed by Unified Automation is selected as our OPC UA client application and run on the Windows PC.

4.5.1 Embedded platform. The BBB (BeagleBone Black) [16] is used as the embedded platform to run the OPC UA server implementation from AsNeG. The BBB runs Debian OS, has a 512MB DDR3 RAM, 4GB 8-bit eMMC on-board flash storage and connects to the host PC via its USB host and Ethernet interfaces [16].

The tools required to build and run this implementation on the BBB are as follows:

- Boost version 1.54
- Gcc 4.9
- Cmake
- Openssl
- OPC UA server configuration file (.xml)
- EDDL files (.ddl)
- EDDL configuration files (.xml)

The build and run process of the OPC UA server implementation have been automated by bash scripts. To start the OPC UA server, a configuration file which defines information models, security profiles, transport profiles is required.

At run-time, the OPC UA server is able to load several EDDL configuration files. Each EDDL configuration file contain information about the EDDL file that shall be loaded by the OPC UA server and predefined ids and attributes of the OPC UA nodes that would be created by the OPC UA server. For example, the EDDL configuration file for the NIKI Current Sensor as shown in Figure 5 defines the corresponding EDD description of the NIKI Current Sensor in Figure 2. The OPC UA server therefore loads several EDDL configuration files, extracts the specified EDDL files, invokes the EDDL parser, maps the parsed EDDL languages elements to the corresponding OPC UA nodes as defined in [15] and finally creates the corresponding OPC UA nodes in the address space of the OPC UA server.

```
<?xml version="1.0" encoding="utf-8"?>
<EddlModel xmlns="http://eddl/EddlModel.xsd">
  <EddlPath File="@CONF_DIR@/Nodes/
    Current-Sensor.ddl" />
  <ObjectNode Id="998" BrowseName="NIKI
     Current Sensor" DisplayName="NIKI Current
    Sensor" Description="Current Sensor" />
  <VariableNodes FirstId="1200" />
</EddlModel>
```

Figure 5. EDDL Configuration file for NIKI Current Sensor.

4.5.2 Windows PC. A standard Windows PC is used to host the UaExpert application. UaExpert is a full-featured OPC UA client implementation that supports several OPC UA functionalities such as data access, alarms and conditions and historical access. In our test setup, the UaExpert is used to connect and visualize the OPC UA nodes created on the OPC UA server.

4.6 Test results

Figure 6 shows a snapshot of the UaExpert's visualized OPC UA nodes that were created dynamically by the OPC UA server running on the BBB after loading and parsing sample NIKI Current and Temperature Sensor EDDL configuration files and their corresponding EDD files.

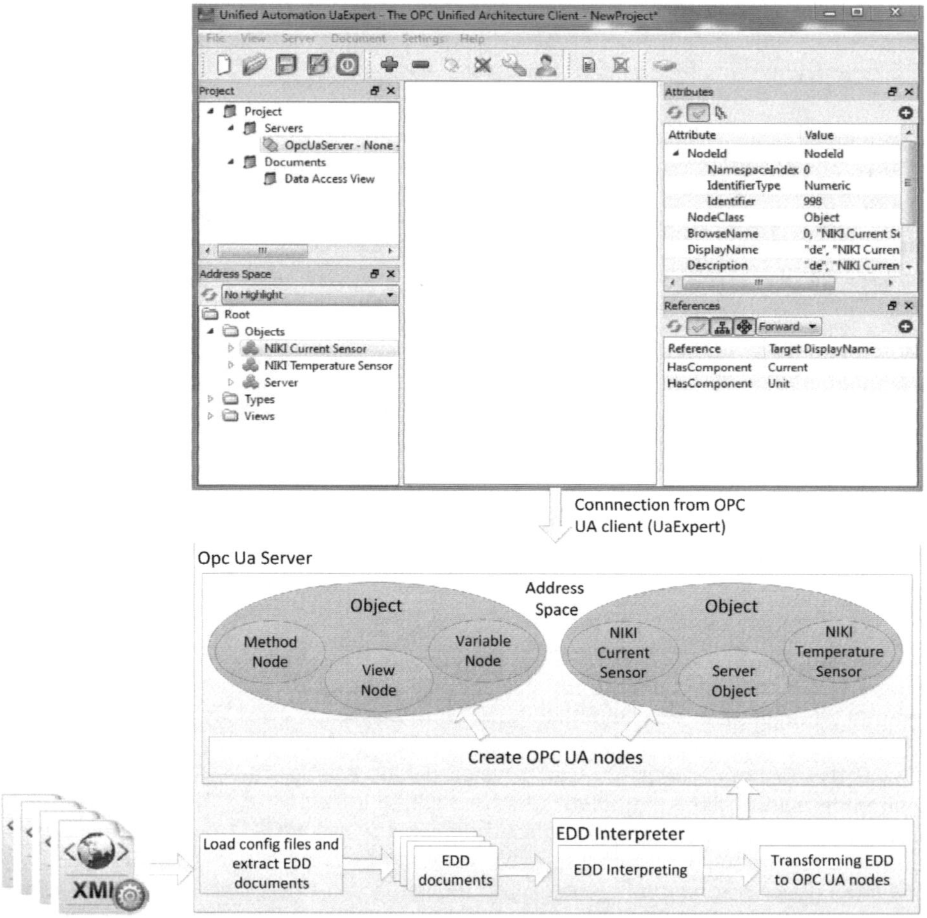

Figure 6. Snapshot showing UaExpert visualized nodes from the OPC UA server.

5. Outlook

We have presented in this paper our approach to dynamically map device descriptions based on EDDL to OPC UA. So far, our test results are based on Profibus, FOUDATION and HART devices. Currently, work is ongoing to integrate the open source LWM2M server implementation in the OPC UA server to allow the OPC UA server to access or get notified of data changes in the corresponding client, for example, a temperature sensor hosting a LWM2M client. Furthermore, we are working to support more devices based on protocols such as Modbus.

References

[1] M. Gunzert. Compatibility and interoperability in field device integration – A view on EDDL, FDT and FDI, (2015). In IEEE Conference Publications pp. 941-946G. Lütjering, J.C. Williams, *Titanium*, Springer, 2003.
[2] S. Runde, G. Wolf, M. Braun. EDDL and Semantic Web – from Field Device Integration (FDI) to Future Device Management (FDM), (2013). In IEEE Conference Publications pp. 1-8.
[3] www.eddl.org.
[4] www.fdtgroup.org.
[5] http://www.fdi-cooperation.com/technology.html.
[6] OPC Foundation, OPC Unified Architecture Specification Part 1: Overview and Concepts, Release 1.03 (2015)
[7] OPC Foundation, OPC Unified Architecture Specification Part 3: Address Space Model, Release 1.03 (2015)
[8] OPC Foundation, OPC Unified Architecture Specification Part 5: Information Model, Release 1.03 (2015)
[9] PROFIBUS Nutzerorganisation e.V., EDDL Specification for Profibus Device Description and Device Integration, (2005).
[10] https://github.com/open62541/open62541/wiki/List-of-Open-Source-OPC-UA-Implementations.
[11] https://github.com/OPCFoundation/UA-.NET
[12] http://open62541.org/
[13] http://wiki.asneg.de/wiki/index.php/Produkte/OpcUaStack
[14] https://81.169.197.52:8443/tree/OpcUaStack.git/master/src
[15] FDI Cooperation, FDI Technical Specification Part 5: Information Model (2015)
[16] https://cdn.sparkfun.com/datasheets/Dev/Beagle/BBB_SRM_C.pd

Acknowledgement

The work of this paper was partially supported within the project NIKI4.0 "NIcht disruptives Kit für die Evalution von Industrie 4.0" funded by Baden-Württemberg Stiftung GmbH. The authors are grateful for this support.

IOP Conf. Series: Journal of Physics: Conf. Series **870** (2017) 012007 doi:10.1088/1742-6596/870/1/012007

Impact of Machine Virtualization on Timing Precision for Performance-critical Tasks

Kirill Karpov[1], Irina Fedotova[1] and Eduard Siemens[1]

[1]Anhalt University of Applied Sciences - Faculty of Electrical, Mechanical and Industrial Engineering, Bernburger Str. 57, 06366, Köthen, Germany

kirill.karpov@hs-anhalt.de

Abstract. In this paper we present a measurement study to characterize the impact of hardware virtualization on basic software timing, as well as on precise sleep operations of an operating system. We investigated how timer hardware is shared among heavily CPU-, I/O- and Network-bound tasks on a virtual machine as well as on the host machine. VMware ESXi and QEMU/KVM have been chosen as commonly used examples of hypervisor- and host-based models. Based on statistical parameters of retrieved distributions, our results provide a very good estimation of timing behavior. It is essential for real-time and performance-critical applications such as image processing or real-time control.

1. Introduction

Modern computer systems are often virtualized to sufficiently increase the hardware utilization, to raise the flexibility of system configuration and to reduce maintenance costs. However these advantages come with several challenges due to the virtualization concept, which can dramatically impact the support of real-time and high precision applications. While the host operating system controls the access to the physical resources directly, the guest OS has to traverse through the host operating system layer before it can access the resources. This additional software layer impacts performance and may be critical for applications which require precise and accurate timing functionality such as video streaming, video games or medical surgery equipment control.

Keeping time precision of control algorithms and meeting soft real-time requirements under virtual environments is a complex problem that requires efficient access to the physical platform [1]. Typically the peripheral hardware of computing system is not directly accessible by the user-level application software layers or even the host side without any form of virtualization. Thus, timing operations provided by the OS are too inefficient for real-time tasks [2]. While on Linux direct access to the TSC counter can improve context switching on Linux by avoiding expensive system calls for time acquisition, precise delay measurements are a more challenging task. In fact, timer access through the RDTSC machine instruction available for x86-64 architecture [3, 4] can cancel out the speed benefits by invoking a call up to 2-3 times. However, a sleep procedure, whose realization in the standard GNU C library has an insufficient precision due to the long wake-up time, can even more aggravated when sharing resources between host and virtual machines (VM).

While applying different techniques for timekeeping on Linux machines, we intend to demonstrate problems that can arise for time-critical applications running in a virtualized environment. Two basic [5] virtualization classes have been studied: the hypervisor-based model (with VMware ESXi) and the hosted-based model (QEMU/KVM). The hyperviser-based model, also known as virtual machine monitor (VMM), is essentially a tiny operating system, which allocates hardware resources among VMs. In contrast, hosted-based implementations use already present operating system resources and together with a virtualization module act as a VMM. In this case, the host OS integrates a VMM layer that is responsible for providing the VMs with their virtual platform interface and managing all context switching and scheduling. Such segregation will obviously add additional overhead in reaction to the timer interrupts and can increase the cost for time measurement. Therefore, considering the difference between communication with the underlying physical hardware and resource management, the quantitative and qualitative analysis of timing behavior on both models is essential for the planning of real-time applications on virtualized systems. Moreover, the comparison should include not only the proof how fast the hardware access from guest operating system level can be, but also how precisely a system wakes up from a sleep operation in a virtual environment.

The remainder of this paper is structured as follows: Section 2 reviews related work. Section 3 is devoted to the experimental setup, describes hardware and software equipment, load types, measurement and representation methods for the current research. The results of experiments with time acquisition and sleep function measurements is presented in Section 4 and 5 respectively. Finally Section 6 discusses the results, followed by the future work and the conclusion in Section 7.

2. Related Work

A detailed performance analysis of Xen Linux was made by P. Barham [6]. The authors made performance measurements in the hosting scenario on the Xen hypervisor-based VMM with different loads to simulate real life servers and compared it with different virtualization software.

A deep analysis of influence of visualization on network performance on Amazon EC2 service was performed in [7]. The authors have considered a wide number of network parameters such as Bandwith, RTT, delay, packet loss rate, throughput and provided early results towards understanding the characteristic of Xen-based virtualized data centers.

S. G. Langer and T. French [8] have conducted research of influence of virtualization on I/O operations. It has been performed on Red Hat, Xen, ESXi, kvm_redhat64 as hypervisors, VMWare player and VirtualBox as hosted-based VMMs.

Moreover, a few studies have investigated the impact of machine virtualization on time related functionality. For instance, authors in [9] have done a comprehensive analysis of different time acquisition methods. They simulated idle, CPU and I/O-loaded systems on a host machine and partly compared the results on five different virtual platforms: VMware Player, VirtualBox, QEMU, Xen, and KVM. However, they performed this without observation of sleep precision.

The innovation novelty of our work is the full investigation of virtualized timer hardware, which includes not only measurements for time acquisition, but also the reaction of timers on waiting operations. Moreover in this work, we intended to create conditions as close as possible to real world applications running on virtual systems, simulating different kind of loads on both host and guest sides.

3. Experimental setup

3.1. System configuration

All tests are performed on the same hardware platform, a simple hardware server-like set with a main requirement to CPU to include Intel-VT/AMD-V Virtualization Technology (VT) support [10]. For the test Intel(R) Core(TM) i7-860 @ 2.80GHz (Lynnfield) with 4 physical cores along with 8 GB of RAM and Broadcom BCM5751 Netxtreme Gigabit NIC have been used. All measurements on both host and guest operating systems have been performed on Ubuntu 14.04.4 LTS 4.2.0-34-generic kernel with standard scheduling configurations. Measurements have been performed on the system with invariant TSC counter with a constant rate. Hyper-threading mode has been disabled to derive a serialized execution. Virtual machines have been configured with enabled VT technology to replicate physical hardware, and were provided with two CPUs and 1 GB of RAM each.

IOP Conf. Series: Journal of Physics: Conf. Series **870** (2017) 012007 doi:10.1088/1742-6596/870/1/012007

3.2. Load types
In the context of multiprogrammed computing system, it's possible to distinguish several main types of tasks [11, 12]. According to this classification, we distinguish the following load scenarios and create corresponding testbeds for the experiment:

Idle testbed – in this load type operating system works in normal mode, without any additional working applications. An idle system represents the best-case scenario.

CPU testbed – this type creates very large CPU load using up to 100 % of available CPU resources. The stress utility creates 100 threads with square root calculations. This type of load is often created by using any video/audio conversion/compresion algorithms or applications for graphic processing.

IO/Network testbed – this load type simulates typical VMs load, which can exist on hosting in a large data center. It creates NFS server on VM and starts sending and receiving data calling standard read/write calls. This workload results in a large number of context switches and disk interrupts. As an example, this scenario can be used for solving such problems like distributed processing of big data.
Network testbed – in this mode a VM instance runs a Reliable Multi Destination Transport Protocol (RMDT) [13], which runs in bi-directional transmission mode. For the current testbed RMDT has been configured to send repeatedly the same chunks of data from a buffer to avoid stalling on disk IO. The protocol fills all available bandwidth, converging to the limit of network internet card and allows to create an intensive network load in long fat pipes manner.

3.3. Measurement Methodology
First, the experiment evaluates the cost of setting a timer: the time required to obtain a time value from the timing hardware. We use the term *timer cost* as a measuring parameter. Second, the measurements of sleeping operations have been performed and the difference between the targeted wake-up time and actual times of wake-ups (refereed as a *miss time*) have been calculated. For both cases of measuring both the *timer cost* and *miss time*, the usage of system call is contrasted to gathering counter cycles from timer hardware. As an interface to available time counters the HighPerTimer library has been applied [14]. It is used to obtain access to the TSC timer on the current processor from user-space and provides a mechanism for precise sleep operations. The sleep function from HighPerTimer library (*HPTSleep*) allows to significantly decrease the overhead of miss time without starving the CPU.

In present research, the median has been chosen as the main statistical significative. Since the data follows a non-parametrical distribution, the median value carries more information about distribution than the mean value. In the next sections we use two types of plots: plots with raw data with bold lines which are the medians and plots with the representation of Complementary Cumulative Distribution Function (CCDF) with dashed lines which are the means. CCDF is a distribution function that shows how often the random variable is above a particular level. Additionally, the mean values are illustrated within CCDF plots, which allow to estimate how mean values are deviated from medians.

4. Experimental Results For Time Measurements
Time measurements are performed with TCS timers and clock_gettime() system call in three scenarios: Host OS, where operating system is placed on bare hardware, VMware ESXi for hypervisor-based virtualization and QEMU for virtual monitor based virtualization. When acquiring TSC timer data through the RDTSC machine instruction, we will henceforth refer to it as the TSC timer method. Similarly, time acquisitions using the clock_gettime() system call is referred to as system call.

4.1. Idle testbed
Initially all measurements have been performed in idle state in order to compare the results with heavy loads on the next stages. As shown in figure [1], median values on all three environments are almost the same with TSC timers as well as with system calls. In all cases TSC timers show better performance than system call. As expected, the host OS has the lowest time overhead (figure [1] (a)), about 17.2 nsec when using the TCS timer and 27 nsec for the system call. VMware ESXi (figure [1] (c)) shows lower precision than the host OS as well as a deviation and at the same the time mean value is lower than with

QEMU, 20.1 nsec for TSC timers and 34 nsec for the system call. TSC timers on QEMU virtualized OS (figure [1] (b)) have a median value near to the idle value, 21.5 nsec, however the system call has a significantly larger overhead, 66 nsec. Figure [2] shows that data from all sources have quite similar distributions. Comparing results of forwarding of the RDTSC instruction on VMware ESXi and QEMU, VMware shows less overhead on the translation. Considering the CCDFs of giving results that presented in figure [2], it can be established that TSC timer without load can guarantee overhead up to 20 ns with probability 0.999. Also blue and red lines on the plot which correspond to QEMU are grouped together, they have the longest tails in worst case scenario.

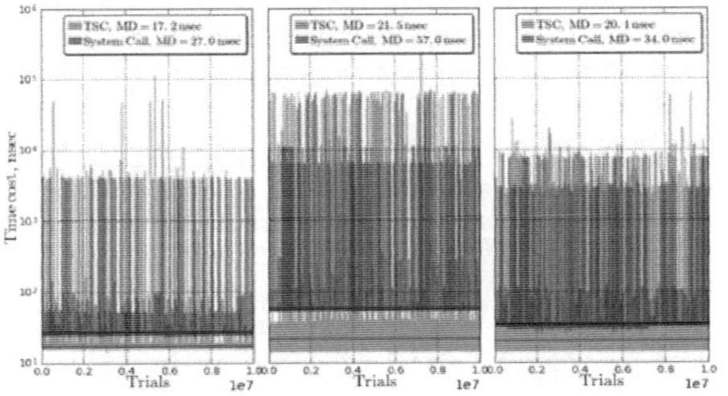

Figure 1: Measurements of timer cost for the Idle testbed on: (a) Host OS (b) QEMU (c) VMware ESXi

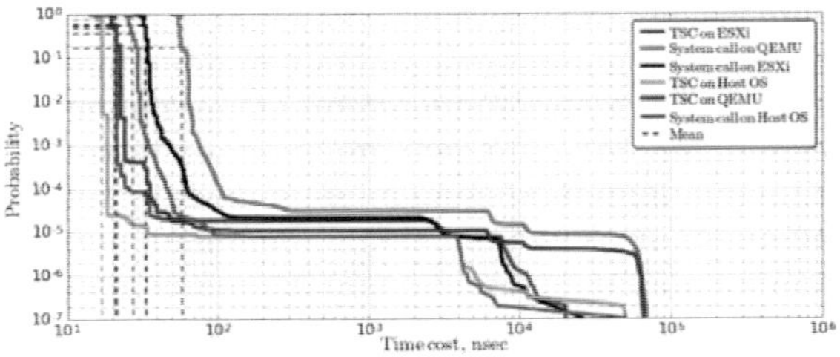

Figure 2: CCDF representation of measurements of timer cost for the idle testbed

4.2. CPU-bound Load

CPU-bound background workload can be expected to result in worsened latencies due to limited hardware resources such as memory bandwidth and shared caches [11]. The results of the experiment with CPU-bound load are shown in figures [3] and [4]. They show that at there is no big effect on median values. However it causes long tails in the CCDF plot, which implies significant influence on the mean of the distributions. The graphs are shifted to the right, to the region of 1 μs. In this experiment, we can establish that probability of exceeding out 20 nsec is lower than 10^{-3} with TSC timer. In other words, we can guarantee that time overhead will be not greater than 20 nsec in 99.99 % of all cases.

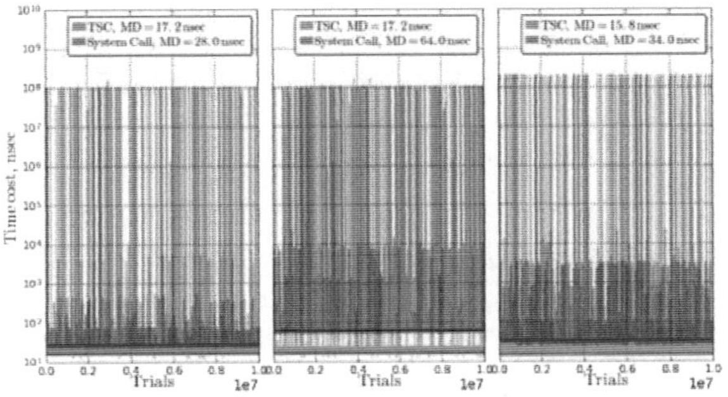

Figure 3: Measurements of timer cost for the CPU testbed on: (a) Host (b) QEMU (c) VMware ESXi

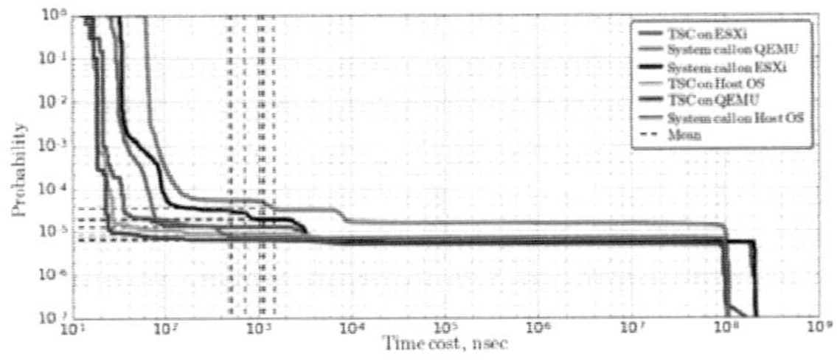

Figure 4: CCDF representation of measurements of timer cost for the CPU testbed

4.3. IO/Network-bound Load

IOP Conf. Series: Journal of Physics: Conf. Series **870** (2017) 012007 doi:10.1088/1742-6596/870/1/012007

Further, challenges of interrupt handling caused by IO-bound tasks and network load are illustrated in figures 5 and 6. Gathering values from TSC shows almost native overhead on all platforms. Although VMware ESXi has median values (figure 5 (c)) of data gathered with TSC timer higher than Host OS and QEMU, it has significantly lower spread than QEMU as well as Host OS. This also can be seen in figure 6, the tails of distributions from VMware ESXi datasets indicate lower time cost value than in other cases. Nevertheless, the virtualized TSC timer on QEMU shows an unexpected advantage: its median overhead value even lower than median value obtained on the same non-virtualized hardware where QEMU placed on. This can be seen on CCDF plot (figure 6 blue line), 90% of outcomes are concentrated below 17 nsec and past this value the distribution line falls steeply and then follows the behavior of other TSC distributions. Also with this plot the region of reliable operation can be determined. Under such load TSC can guarantee that 99.999% of outcomes will not exceed 21 nsec.

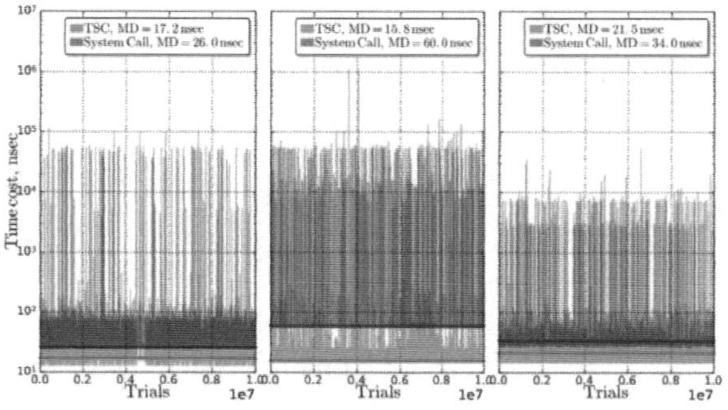

Figure 5: Measurements of timer cost for the IO/Network testbed on: (a) Host (b) QEMU (c) VMware ESXi

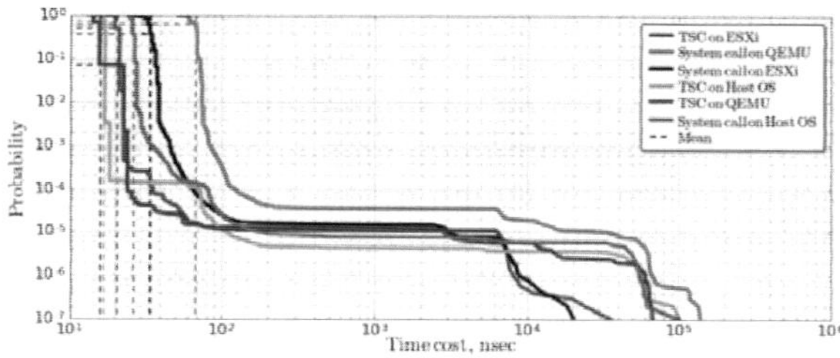

Figure 6: CCDF representation of measurements of timer cost for the IO/Network testbed

4.4. Network-bound Load

The results of the given experiment are presented in figures 7 and 8. In this case, like in previous experiments, the TSC timers show lower time overhead than the system call. However, as seen in figure 7, similarities in measurement results are shared mostly between HPTimer and the system call, rather than between different virtualization environments. Figure 8 shows that these similarities mostly concern the tails of the distributions. Long and heavy tails of distributions that have been obtained with system calls and the TSC counter are grouped together according to the platform. Although distributions with VMware ESXi have the longest tails, TSC timer shows the lowest median value and system call has median value lower than system call on QEMU. Also, based on CCDF plots we can estimate that 99.99 % of all outcomes, obtained with TSC timer during the experiment have values less than 22 nsec.

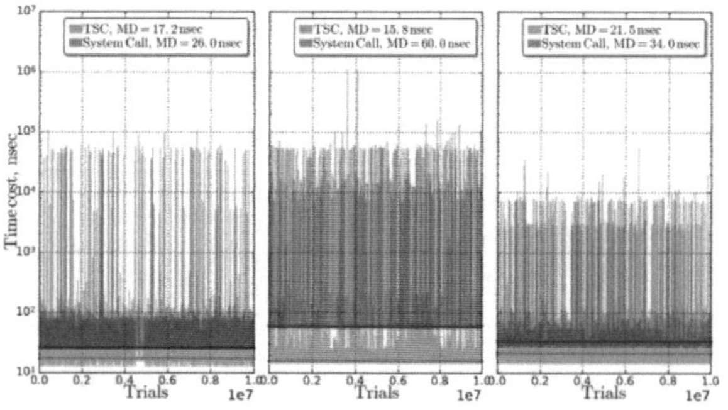

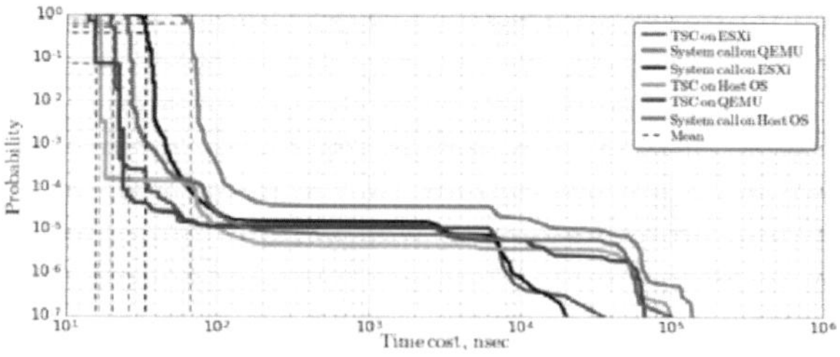

Figure 7: Measurements of timer cost for the Network testbed on: (a) Host (b) QEMU (c) VMware ESXi

Figure 8: CCDF representation of measurements of timer cost for the Network testbed

5. Experimental Results For Sleep Measurements

Sleep measurements are also performed by comparing the sleep function of the HighPerTimer library and the usleep() call of the standard C library in three scenarios: Host OS, VMware ESXi and QEMU. Experiments with the sleep operation from HighPerTimer library are called *HPTSleep* in this paper. Experiments with usleep() form standard C library are called *uSleep*.

5.1. Idle testbed

As in the previous section, measurements have to be performed in idle state in order to establish a baseline when comparing load scenarios. As shown in figure 9, median values during HPTSleep function on all three platforms are in the same range-between 80 and 116 nsec. The invocation of uSleep system call shows miss times that are almost three orders of magnitude greater than HPTSleep function as expected: 60 μsec compared to 80 nsec on Host OS, 97 μsec compared with 116 nsec on QEMU and 63.1 μsec compared with 95 nsec on VMware. Figure 10 shows that the respective distributions of the data are grouped together by the chosen sleep method.

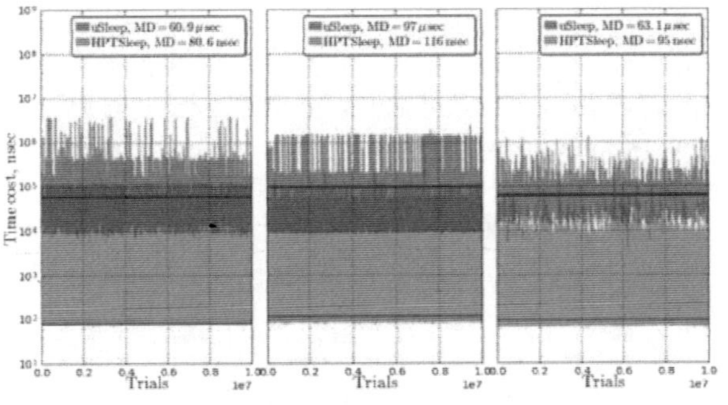

Figure 9: Measurements of miss time for the Idle testbed on: (a) Host (b) QEMU (c) VMware ESXi

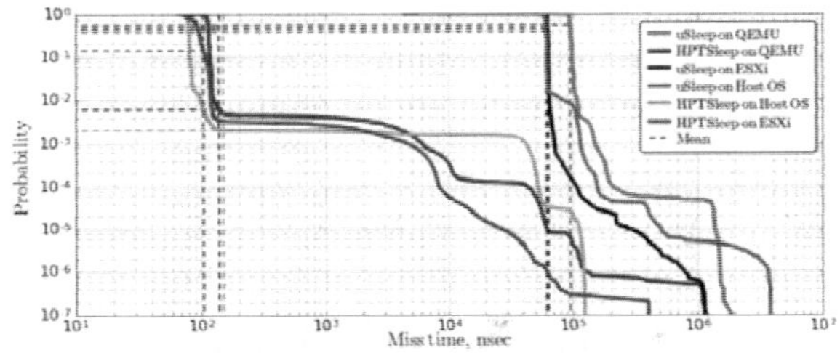

Figure 10: CCDF representation of measurements of miss time for the Idle testbed

5.2. CPU-bound Load

As in the idle scenario, with CPU load the distributions (figure 11) of HPTSleep miss as well as their medians are in the same range on all three platforms: Host OS has a 89.5 nsec median value of miss time (figure 11 (a)), QEMU virtualized OS has 104 nsec (figure 11 (b)), and VMware ESXi has 114 nsec (figure 11 (c)) respectively. Distributions of data with uSleep function are also close to each other: 54.7 μsec on Host OS, 61.3 μsec on QEMU virtualized OS and 56 μsec on VMware ESXi hypervisor-based OS. Considering the CCDF in figure 12, HPTSleep produces much longer tails than uSleep. Mean values are grouped together in the range between 0.1 and 1 msec while median values are around 100 nsec in case of HPTSleep and 50 μsec in case of uSleep. However, median values of HPTSleep differ by 15% maxwhile medians of uSleep function differ up to 45%. Also the CCDF shows that outcomes of the HPTSleep function have a miss time overhead below 120 nsec with a 99.9% probability.

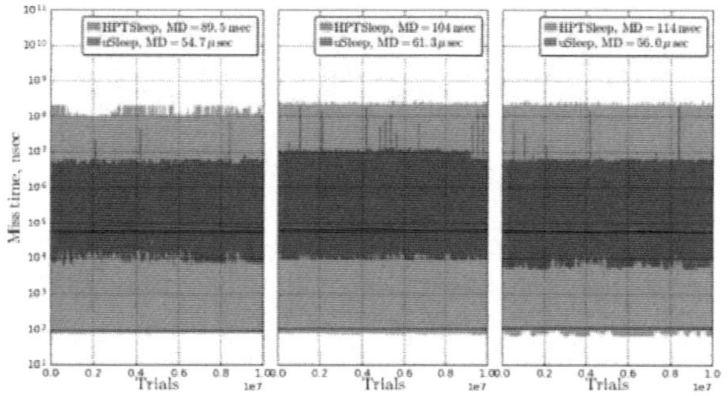

Figure 11: Measurements of miss time for the CPU testbed on: (a) Host (b) QEMU (c) VMware ESXi

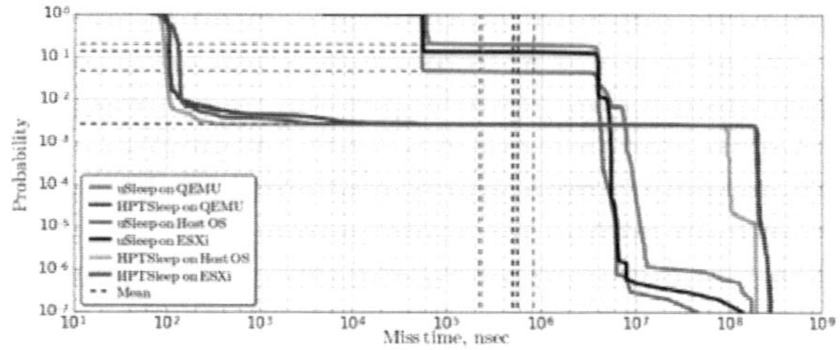

Figure 12: CCDF representation of measurements of miss time for the CPU testbed

5.3. IO/Network-bound Load

The results of this experiment show that although medians of distributions sleep misses with *HPTSleep* are significantly lower than with uSleep function (tens of nanoseconds with *HPTSleep* versus tens of microseconds with *uSleep*) standard deviations of *uSleep* distributions is one order of maginatude lower than HPTSleep. All examined platforms exhibit this behavior. VMware ESXi shows a lower spread than other platforms in case of TSC counter (figure 13 (c)). QEMU has the highest spread with TSC counter within all other platforms (figure 13 (b)). The CCDF plot, in figure 14 shows that all mean values except QEMUs *HPTSleep* mean are located near their medians. This outstanding behavior is caused by rare and powerful outbreaks which are in the same order of magnitude as worst case values for *usleep*. *HPTSleep* function have *miss time* value below in 99% cases.

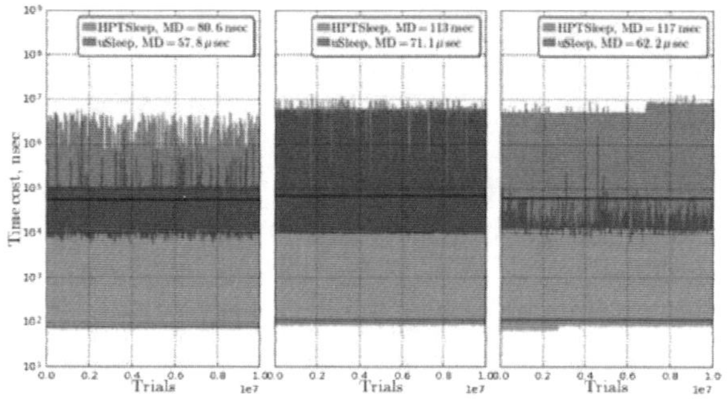

Figure 13: Measurements of miss time for the IO/Network testbed on: (a) Host (b) QEMU (c) VMware ESXi

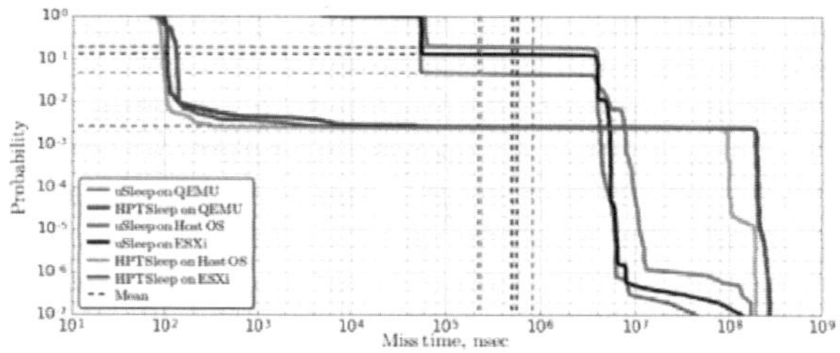

Figure 14: CCDF representation of measurements of miss time for the IO/Network testbed

5.4. Network-bound Load

Similarly to time acquisition measurements with network-bound load, the datasets presented in figures 15 and 16 also have "step-function-like" behavior. This behavior can be explained by fickleness of network data transmission, since the rmdt-protocol tries to take full advantage of available network and CPU resources which could be a cause of network congestions and packet loss that spawns fickleness. However, the CCDF representation (figure 16) is more similar to the IO/Network-bound case. With the system sleep call, mean values of the distribution are near the medians, at the same time in the case of *HPTSleep* they are shifted to the right. With Network-bound load, *HPTSleep* function has *miss time* values below 120 nsec in 99% of cases.

IOP Conf. Series: Journal of Physics: Conf. Series **870** (2017) 012007 doi:10.1088/1742-6596/870/1/012007

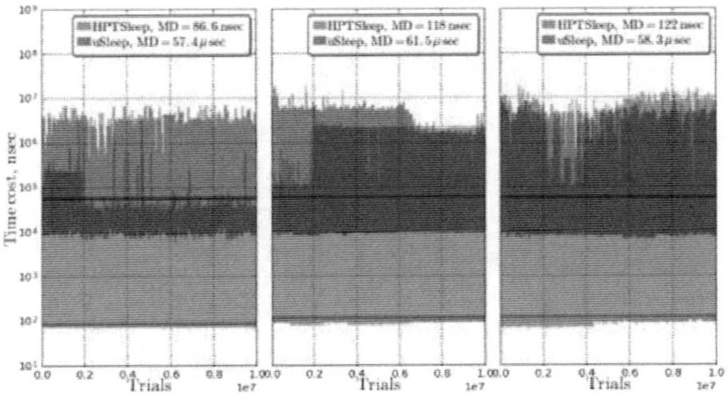

Figure 15: Measurements of miss time for the Network testbed on: (a) Host (b) QEMU (c) VMware ESXi

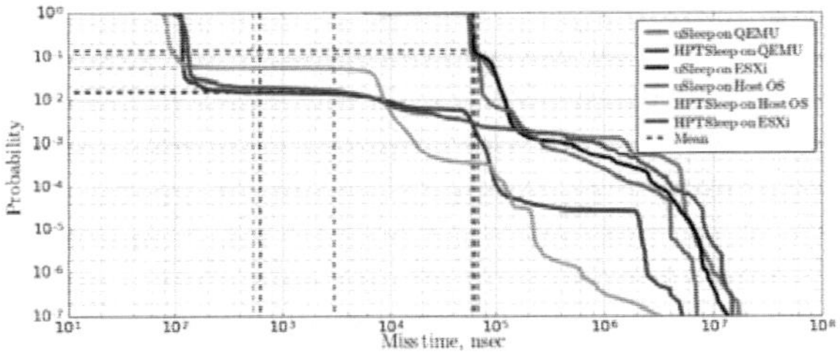

Figure 16: CCDF representation of measurements of miss time for the Network testbed

6. Discussion of Results

In the preceding sections we described in detail the results of our timing measurements for each scenario. However, the cross comparison of distribution patterns for different load types, methods and VM models is required for specifying explicitly the impact of virtualization on timing.

6.1. Loads

CPU-bound load is characterized by very long outliers and as a consequence the mean values are shifted to the right relative to the medians. Median values under this load differ from the idle case only up to 15% in case of HPTSleep and up to 45% in case of *uSleep* function. Under IO/Network-bound load, the patterns of the timing distributions are similar to the idle scenario pattern. Medians, means and tail limits are very close to values in the idle scenario. Nevertheless, on VMware ESXi this load produces tails of distributions which are shorter than in non-virtualized measurements. Moreover, the median value of measurements with TSC counter on QEMU is lower than in the not virtualized scenario without any load. Pure network-bound load produces fickle datasets, due to the dynamic nature of data transmission. These distributions have the second longest tails after CPU-bound load. In this case distributions are strongly related on the virtual environment. This can be deducted from the tails behavior. They are grouped together by the system where measurements have been done. Considering worst-case execution time for applications running real-time, CPU-, IO-bound or network tasks, the difference in tails behavior should be taken into account.

6.2. HighPerTimer vs Standard Linux Calls

It's worth noting that in all scenarios, under every load the improved timing methods of the HighPerTimer library result in a significant increase in precision. Time acquisition operations without load have a lower overhead with direct access to hardware than system call: median values on 45% lower on Host OS, on 97% lower in case of QEMU, and on 47% lower on VMware ESXi. Moreover, using a standard system call for sleep operations can lead to even bigger performance degradations. The difference between the two methods can be up to 1000 times. This can be observed well on all CCDF plots, where distributions are grouped into two branches according to the applied operation.

6.3. Qemu vs VMware ESXi

Since hypervisor based model VMware ESXi has no unnecessary layer between hardware and VMM, a better timing precision was expected. In fact, comparing with QEMU virtualization the overhead of VMware ESXi on timing operations is lower. Nevertheless, it is also close to the overhead of the Host OS, however in case of time acquisition with system call, the difference between QEMU and VMware ESXi or Host OS can be up to 80%.

IOP Conf. Series: Journal of Physics: Conf. Series **870** (2017) 012007 doi:10.1088/1742-6596/870/1/012007

7. Conclusion

Our various measurement results for the case where one virtual instance is located on hardware allow to conclude the following: (i) While VMware ESXi has almost no basis overhead, QEMUs access to the hardware adds relatively high overhead (especially by introducing jitter when waking up from a sleep call); (ii) in case of sleep accuracy, the means of sleeps methods prevail over the virtualization environment effects; (iii) while considering the worst-case execution time, the virtualization used plays an almost neglecting role, the majority of values are measurably better on the host machine than (at least) on QEMU; (iv) the hypervisor-based approach brings clearly measurable benefits compared with the OS-based one.

Our ongoing work is to scale this experiment on the hosting scenario with several VMs and test under additional loads. This would allow us to demonstrate how the time hardware is shared among VMs and identify the resource limits.

Acknowledgment

This work has been funded by Volkswagen Foundation for trilateral partnership between scholars and scientists from Ukraine, Russia and Germany within the project CloudBDT: Algorithms and Methods for Big Data Transport in Cloud Environments.

References

[1] García-Valls M, Cucinotta, T , Lu, C, 2014. Challenges in real-time virtualization and predictable cloud computing. *Journal of Systems Architecture* **60** pp 726 - 40.
[2] I. Fedotova and E. Siemens 2014 *System Time Issues for the ARM Cortex A8 Processor* Proc. of the International Conference on Applied Innovations in IT (Köthen) p 7
[3] Intel 2016 *Intel 64 and IA-32 Architectures Software Developer's Manual* vol. 2
[4] G. Paoloni 2010 *How to benchmark code execution times on Intel IA-32 and IA-64 instruction set architectures* (Intel Corporation) vol 123
[5] Rosenblum, M Garfinkel, T 2005 Virtual machine monitors: current technology and future trends *Computer* **38** pp 39 - 47.
[6] P. Barham et al., *Xen and the Art of Virtualization* in Proc. of the 19th ACM Symposium on OS Principles (New York) pp 164-77
[7] Wang, G, Ng, T S E, 2010 *The Impact of Virtualization on Network Performance of Amazon EC2 Data Center* (San Diego: IEEE)
[8] Langer S G, French, T, 2011. Virtual Machine Performance Benchmarking. *Journal of Digital Imaging*, **24**, pp.883 - 9.
[9] Adamczyk B, Chydzinski A., 2015. Achieving High Resolution Timer Events in Virtualized Environment *I. McLoughlin, ed. PLOS ONE* **10** p.e0130887.
[10] R. Uhlig 2005 Intel virtualization technology *Computer* **38** pp 48–56.
[11] Cerqueira F and Brandenburg B 2013 *A Comparison of Scheduling Latency in Linux, PREEMPT-RT, and LITMUS^{RT}* in 9th Annual Workshop on OSs Platforms for Embedded Real-Time Applications (Paris) pp 19-29
[12] Litayem N, Ben Saoud S, 2011. Impact of the Linux Real-time Enhancements on the System Performances for Multi-core Intel Architectures. *International Journal of Computer Applications*, **17** pp 17 - 23.
[13] Bakharev A V, Siemens E, Shuvalov V P 2014 *Analysis of performance issues in point-to-multipoint data transport for big data* 2014 12th International Conference on Actual Problems of Electronics Instrument Engineering (Novosibirsk: IEEE) p 431
[14] Fedotova I Siemens E H Hu 2013 A high-precision time handling library *Journal of Communication and Computer* **10** pp 1076-86

The detection of dissolved gases in transformer oil by gas chromatography with helium ionization detector

DENG Xian-qin[1], FANG Hua[2], LI Min-xian[1]

1.State Grid Shanghai Electric Power Research Institute, Shanghai, 200437, China;
2.Shanghai Huaai Chromatography Analysis Co.Ltd, Shanghai, 200437, China

Abstract. The GC-PDD with the technology of valve cutting and helium ionization detector was used to analyze the dissolved gases in ultra-high voltage(UHV) and extra-high voltage(EHV) transformer oil. The detection limit(DL) reached ppb grade, especially for the featuring gas—C_2H_2 and H_2,whose DL could reach 5ppb and 11ppb respectively.The test reproducibility of the instrument was about 1% and the correlation coefficient of standard curve-r is greater or equal to 0.99,which showed obvious advantage compared with normal GC. In addition, the auxiliary gas of H_2 was not used in this instrument, which completely improved the safety performance. Thus, the application of GC-PDD has significant meaning in warning potential malfunction inside the ultra-high voltage transformer in advance.

1. Introduction

With the continuous development and progress of power system, ultra-high voltage transmission of electricity is becoming more and more extensive. And the safe and reliable operation of UHV and EHV transformer equipment plays a very important role in the reliability of urban power supply system. At present, a large number of domestic EHV transformers are easy to cause larger accidents due to higher voltage, micro-discharge and overheating, so that higher requirements are put forward for prevention and detection of early accidents of dissolved gases in transformer oil. So the requirement for instrument detection limit is higher, the application of traditional chromatography is unable to meet the requirements for analysis[1].

Dissolved gases in the transformer oil include hydrogen, methane, ethane, ethylene, acetylene, carbon monoxideand carbon dioxide. In the early stage of fault, the gas formed is dissolved in oil; when the energy of fault is large, it may also gather into free gas [2,3]. Therefore, it is of great significance to periodically measure the composition and content of gases dissolved in oil for the early detection of the potential faults inside oil-filled power equipment. In order to avoid the misunderstanding of electrical equipment failure, IEC 60567[4] and IEC 60599[5] had put forward specific requirements for the test cycle and the detection methods.

In current IEC standard, dissolved gases had been separated from the oil by headspace gas taking or shock separation. In the IEC detection method, two kinds of detectors and a chemical conversion device should be applied. Han Yuwang had detected dissolved gas with AC pulse helium ionization detector ,which proofed the applicability of this detector [6]. J. Jalbert, R. Gilber, and P. Te' treault have ever detected the dissolved gases by headspace degassing method with helium ion-detector gas chromatograph and proofed a higher detection limit and enhanced space [7]. In this paper, helium ionization detector was chosen too. Helium ionization detector is a universal detector, especially for especially for trace gases. The helium ion gas chromatograph was proposed, which can not only simplify the chromatographic process but also improve detection sensitivity.

IOP Conf. Series: Journal of Physics: Conf. Series **870** (2017) 012008 doi:10.1088/1742-6596/870/1/012008

2. Experimental part

2.1. New GC-9560-PDD helium ionization gas chromatograph

Hereinafter referred to as GC-9560-PDD; Shanghai Hua'ai Chromatography Analysis Co. Ltd is equipped with helium purifier and helium ionization detector and its process is converted to double 10-way valve switching process as shown in Figure 1. Injector used is a packed column injector; carrier gas is purged by control of on-off valve. Nitrogen is used as valve-driving gas and usual gas taking port design like ordinary transformer oil instrument is used as gas taking port. Four-way valve seal the system is adopted.

2.2. Traditional GC-9560-HD gas chromatograph

Hereinafter referred to as GC-9560-HD; Shanghai Hua'ai Chromatography Analysis Co. Ltd is equipped with thermal conductivity detector and dual hydrogen flame detector with the process of IEC three detectors as shown in Figure 2.

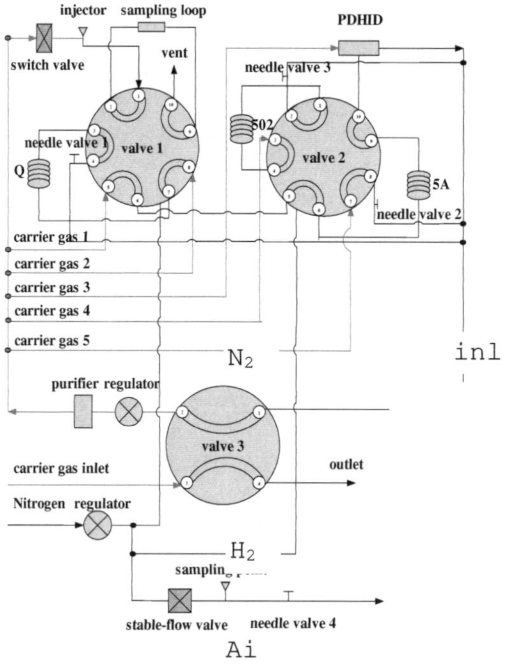

Figure 1. Flow chart of GC-9560-PDD

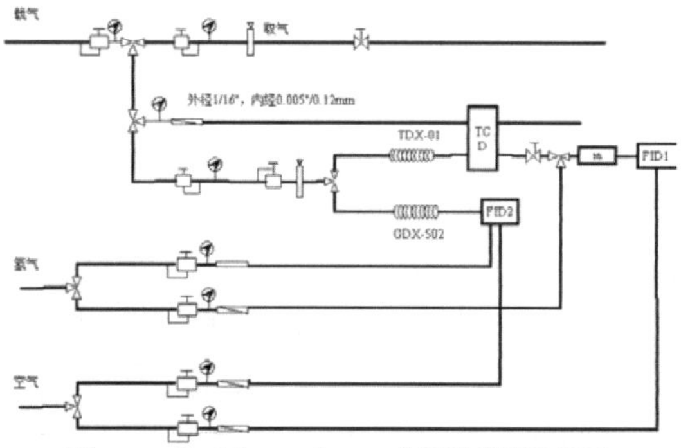

Figure 2. Flow chart of GC-9560-HD

IOP Conf. Series: Journal of Physics: Conf. Series **870** (2017) 012008 doi:10.1088/1742-6596/870/1/012008

2.3. Standard gases

Standard gases (certificate number:15-09680) was Made by Shanghai Weichuang Standard Gas Co., Ltd. The density value of standard gas was shown in Table1. High-purity helium （99.999%） and High-purity nitrogen （99.999%） was made by Shanghai Pujiang Gas Co., Ltd.

Table 1. The component of standard gas.

Component	Content （10^{-6}V/V）	Component	Content （10^{-6}V/V）
CH_4	101	C_2H_2	50.0
C_2H_6	99.4	H_2	985
C_2H_4	99.3	CO	1014
CO_2	2993	N_2	rest

3. Results and discussion

3.1. Flow of new instrument

As shown in Figure 1, GC-9560-PDD has gone through a large number of experiments and continuous improvement and is pre-division of column Q, which guides hydrogen, oxygen, nitrogen, methane and carbon monoxide into column 5A. Column 5A will separate hydrogen, oxygen, nitrogen, methane and carbon monoxide, in which, hydrogen peak is first-out before switching to another column GDX-502, carbon dioxide, ethylene, ethane and acetylene emerge then column Q, and detector switches to 5A, this moment, oxygen and nitrogen have been vented before methane and carbon monoxide peaks emerge. It takes 7 minutes for one injection of spectrogram. Figure 3 is a typical spectrogram of standard gas.

3.2. Comparison of new instrument with IEC process

Figure 4 is a chromatogram of GC-9560-HD. The chromatographic instrument used three detectors, included a thermal conductivity detector and two hydrogen flame ionization detectors. Thermal conductivity detector is used to detect hydrogen, hydrogen flame ionization detector to detect carbon monoxide and carbon dioxide which are converted to methane by reformer, and another way to detect the hydrocarbons such as methane, ethane, ethylene and acetylene. This method wants more detectors, in which hydrogen is used as auxiliary gas; in case of explosion, risk is higher. And the new GC-9560-PDD only needs helium as carrier gas, with nitrogen gas as auxiliary gas so as to greatly improve safety because both the gases are inert gases.

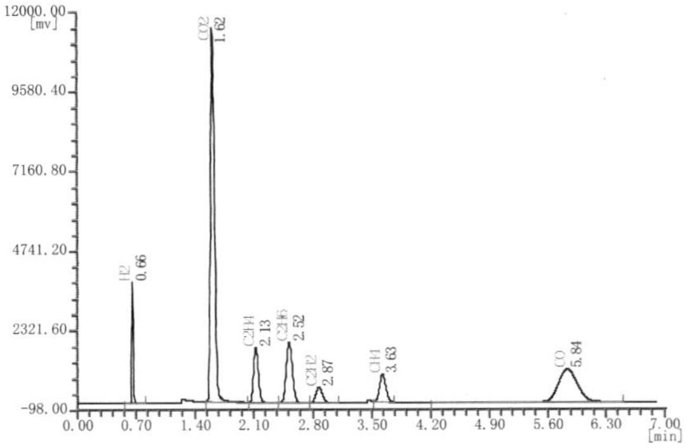

Figure 3. Typical chromatogram of standard gas by GC-9560-PDD

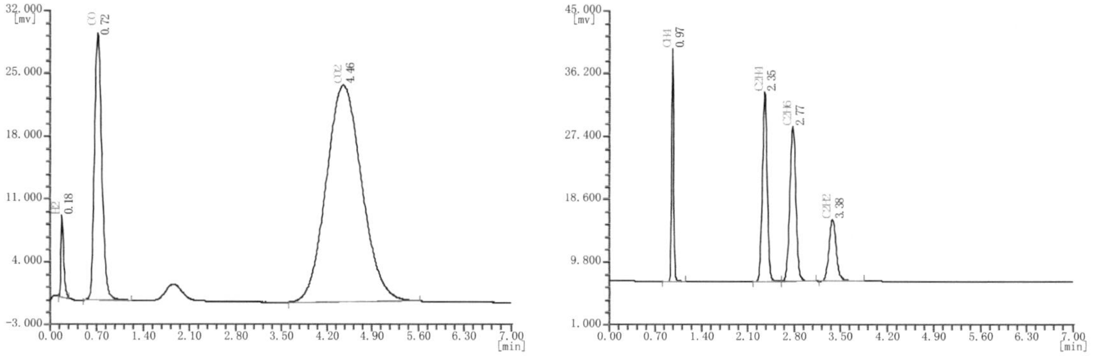

Figure 4. Typical chromatogram of standard gas by GC-9560-HD

3.3. Precision of the new instrument

Table 2 shows the repeatability of GC-9560-PDD in measuring the peak area of standard gases. Test repeatability of the instrument was less than 1%. It was indicated that the precision of the method satisfy test requirement.

3.4. Detection limit of seven gases on the new instrument

The theoretical detection limit of chromatograph was known as the short baseline noise of the instrument within 30 minutes, and the baseline noise is taken as the detection limit of 3 times, as shown in Formula1.

$$D = \frac{3 \times N \times Q}{I} \qquad (1)$$

D - detection limit, N - noise, Q - sample size, I - signal response value,

Table 3 shows the comparison between BG detection limit, the theoretical detection limit (hereinafter known as DL) of GC-9569-HD and GC-9560-PDD. It can be found that GC-9560-HD meet IEC 60599, the theoretical DL of GC-9560-PDD is nearly 80 times higher than GC-9560-HD, and other gases over 5 times. The DL of C2H2 reached 5 ppb, which gives the greatest impact on fault judging.

Table 2. Repeatablity test of standard gases by GC-9560-PDD

Component	H_2	CO_2	C_2H_4	C_2H_2	C_2H_6	CH_4	CO
1st time	4590615	40941603	6555893	9059229	2507532	4526240	15797063
2nd time	4596476	41202099	6689060	9242900	2572068	4589630	15927514
3rd time	4566840	41020204	6665540	9218908	2563812	4560007	15771018
4th time	4547445	40949858	6653227	9201587	2561423	4521657	15572227
5th time	4534646	40741671	6612642	9136720	2543010	4501460	15583729
6th time	4586165	40873084	6648826	9186312	2561244	4518879	15515989
AVERAGE	4570364	40954753	6637531	9174276	2551515	4536312	15694590
RSD	0.55%	0.37%	0.71%	0.73%	0.92%	0.71%	1.03%

IOP Conf. Series: Journal of Physics: Conf. Series **870** (2017) 012008 doi:10.1088/1742-6596/870/1/012008

Table 3. The comparison of DL

Component	H_2	CH_4	CO_2	C_2H_4	C_2H_2	C_2H_6	CO
DL in IEC 60599	≤2	/	≤25	/	≤0.1	/	≤25
DL of GC-9560-HD (10^{-6}V/V)*	0.881	0.025	0.997	0.03	0.046	0.037	0.273
DL of GC-9560-PDD (10^{-6}V/V)**	0.011	0.005	0.011	0.003	0.005	0.003	0.041

Note: *The noise of the data is 4 μ V
**The noise of this data is 20μV

3.5. Linearity of the new instrument to seven gases

Figure 5 shows the standard curve spectrogam of H2 after dilution of the standard gases of different multiples, a standard curve got at the lowest concentration of 1.97 × 10-6 to 985 × 10-6, with the linear dependency R2 of 0.9965. Figure 6 shows the standard curve spectrogam of CO2 after dilution of the standard gases of different multiples, a standard curve got at the lowest concentration of 5.99×10-6 to 2993×10-6, with the linear dependency R2 of 0.9939. Figure 7 shows the standard curve spectrogam of ethylene after dilution of the standard gases of different multiples, a standard curve got at the lowest concentration of 0.199×10-6 to 99.4×10-6, with the linear dependency R2 of 0.9996. Figure 8 shows the standard curve spectrogam of ethene after dilution of the standard gases of different multiples, a standard curve got at the lowest concentration of 0.199×10-6 to 99.3×10-6, with the linear dependency R2 of 0.9997. Figure 9 shows the standard curve spectrogam of acetylene after dilution of the standard gases of different multiples, a standard curve got at the lowest concentration of 0.10×10-6 to 50.0×10-6, with the linear dependency R2 of 0.9972. Under the low concentrations of 0.1 ppm, peaking is still up to 2900 μV, showing a good peaking at low concentrations. Figure 10 shows the standard curve spectrogam of methane after dilution of the standard gases of different multiples, a standard curve got at the lowest concentration of 0.20×10-6 to 101.0×10-6, with the linear dependency R2 of 0.9970. Figure 11 shows the standard curve spectrogam of carbon monoxide after dilution of the standard gases of different multiples, a standard curve got at the lowest concentration of 2.2×10-6 to1014×10-6, with the linear dependency R2 of 0.9977.

Through the linear experiment, it is found that within 500 times of diluted standard gases, there is still a good linear on the instrument, no way to be diluted to lower concentration, well consistent with the result of headspace method helium ion chromatography made by J. Jalbert, R. Gilber and P. Te' treault, with a good linear [7].

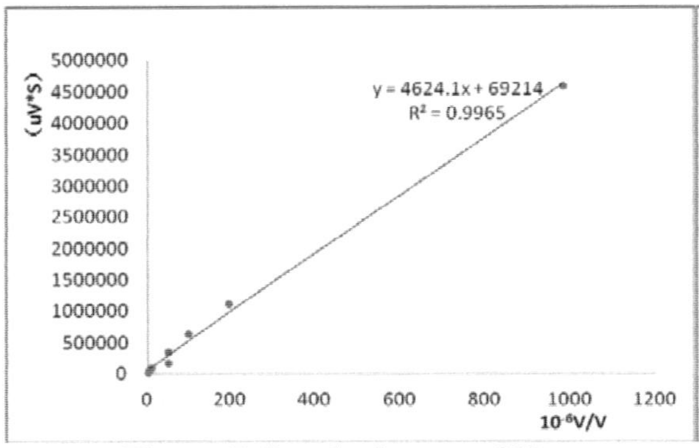

Figure 5. Standard curve of H_2

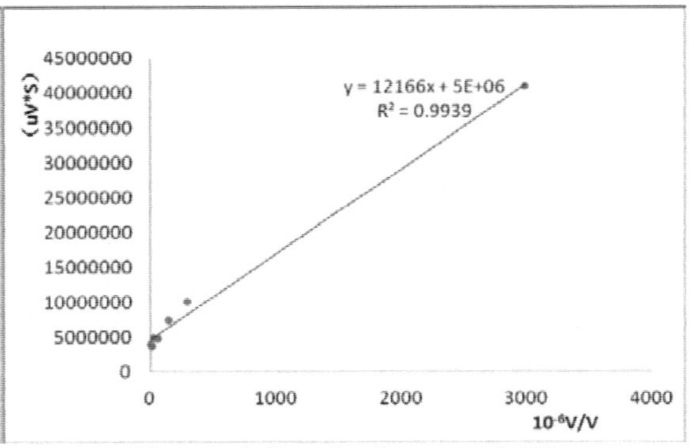

Figure 6. Standard curve of CO_2

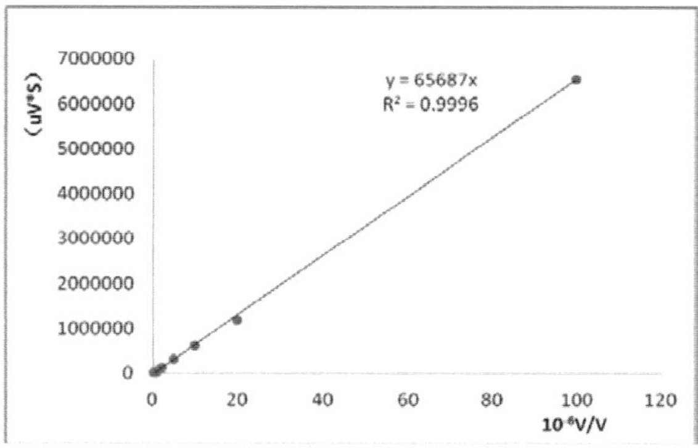

Figure 7. Standard curve of C_2H_4

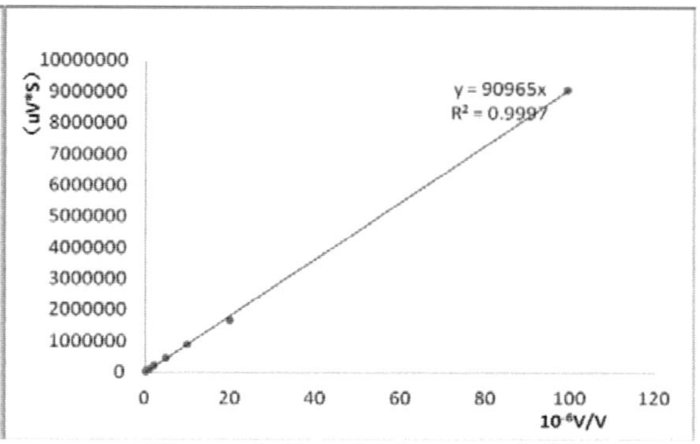

Figure 8. Standard curve of C_2H_6

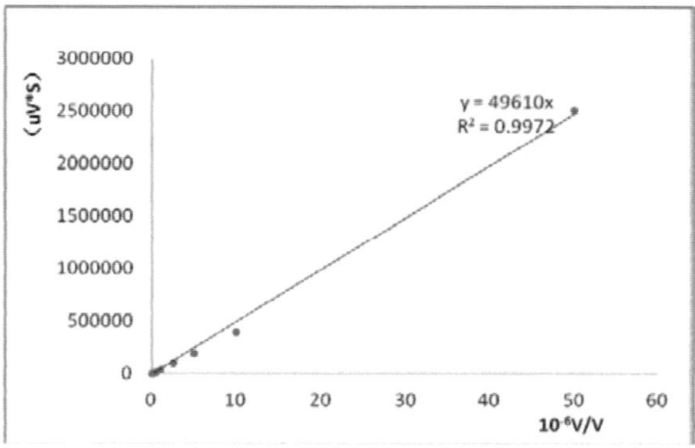

Figure 9. Standard curve of C_2H_2

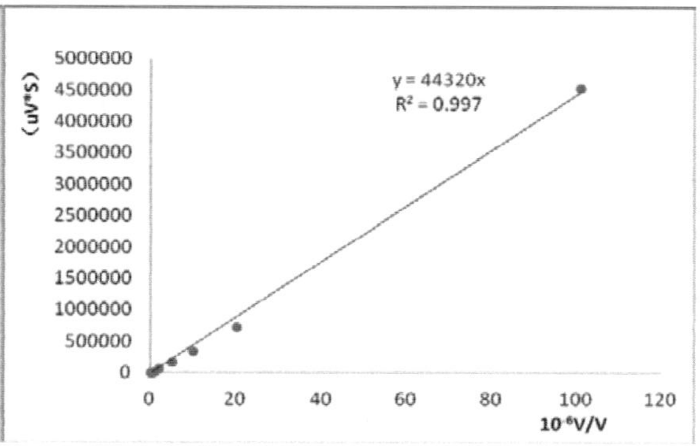

Figure 10. Standard curve of CH_4

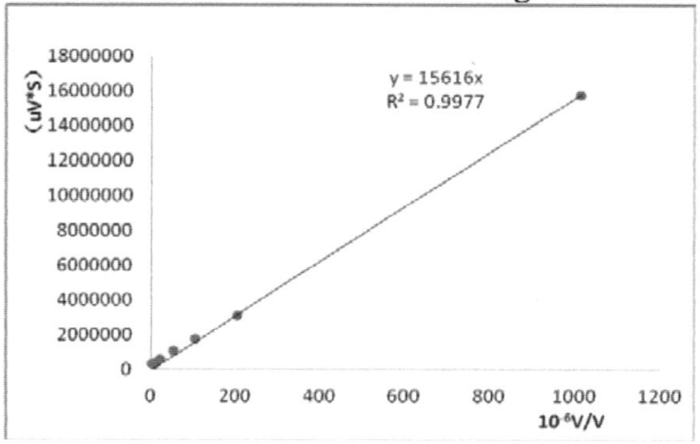

Figure 11. Standard curve of CO

4. Conclusion

A new type of helium ionization gas chromatography with helium ionization detector has been reported for the first time. Test result indicated that it can be used to separate and detect 7 kinds of dissolved gas in transformer oil within 7 minutes. The advanced nature of the instrument is reflected in the detection limit. Detection limit is increased nearly 80 times higher than the detection limit H_2 by traditional GC-9560-HD gas chromatograph, increased over 5 times than other gases. The C_2H_2 which has the greatest impact on the determination of fault has reached a detection limit of 5 ppb. The test reproducibility of the instrument was about 1%. The correlation coefficient of standard curve-r is greater or equal to 0.99. The ultra-low detection limit provided a better platform for the early prevention and fault diagnosis of extra-high voltage transformers. In addition, because of the absence of hydrogen, the safety is quantitatively improved.

IOP Conf. Series: Journal of Physics: Conf. Series **870** (2017) 012008 doi:10.1088/1742-6596/870/1/012008

References

[1] HUANG Jian-bin , LIANG Zhan-yuan , TAN Jian-min ,Method of headspace oil-gas separation degassing for transformers insulation oil[J] . Journal of Electric Power Science and Technology 2013,28(2),89-93.

[2] LIAO Rui-jin, YANG Li-jun, ZHENG Han-bo, et al. Reviews on Oil-Paper Insulation Thermal Aging in Power Transformers[J]. Transactions of China Electrotechnical Society,2012,27(5),1-12.

[3] CHEN We-igen, LIU Bing-jie, ZHOU Heng-yi, et al. The Temperature Characteristics of Photoacoustic Spectroscopy Detection for Dissolved Gases in Transformer Oil[J]. Transactions of China Electrotechnical Society, 2010,25(11),15-20.

[4] Oil-filled electrical equipment - Sampling of gases and analysis of free and dissolved gases - Guidance[S].

[5] Mineral oil-impregnated electrical equipment in service--Guide to the interpretation of dissolved and free gases analysis[S].

[6] HUANG Jian-bin,TAN Jian-min,LIANG Zhan-yuan, et al. Method of headspace oil-gas separation degassing for transformers insulation oil [J]. Journal of Electric Power Science and Technology, 2013,28(2),89-93.

[7] Jalbert J, Gilbert R, Tetreault P. Simultaneous determination of dissolved gases and moisture in mineral insulating oils by static headspace gas chromatography with helium photoionization pulsed discharge detection [J]. Analytical Chemistry, 2001,73, 3382-3391.

IOP Conf. Series: Journal of Physics: Conf. Series **870** (2017) 012009 doi:10.1088/1742-6596/870/1/012009

A method of reconstructing the spatial measurement network by mobile measurement transmitter for shipbuilding

Siyang Guo[1], Jiarui Lin[1,2], Linghui Yang[1] ,Yongjie Ren[1] and Yin Guo

[1] State Key Laboratory of Precision Measuring Technology and Instruments, Tianjin University, Tianjin, 300072, China

gsy135gsy@tju.edu.cn

Abstract. The workshop Measurement Position System (wMPS) is a distributed measurement system which is suitable for the large-scale metrology. However, there are some inevitable measurement problems in the shipbuilding industry, such as the restriction by obstacles and limited measurement range. To deal with these factors, this paper presents a method of reconstructing the spatial measurement network by mobile transmitter. A high-precision coordinate control network with more than six target points is established. The mobile measuring transmitter can be added into the measurement network using this coordinate control network with the spatial resection method. This method reconstructs the measurement network and broadens the measurement scope efficiently. To verify this method, two comparison experiments are designed with the laser tracker as the reference. The results demonstrate that the accuracy of point-to-point length is better than 0.4mm and the accuracy of coordinate measurement is better than 0.6mm.

1. Introduction

With the ongoing process of automation, large-scale metrology [1],[2],[3] plays a vital role in order to ensure product quality in the manufacturing industry. Especially in the shipbuilding industry, accurate and efficient measurement is necessary to provide the dimensional control for a quality product [4]. In traditional shipbuilding industry [5], the dimensional control method with angle square and gradienter is inefficient and inevitably brings the error caused by the human. Nowadays, lots of measurement systems have been applied in the shipbuilding industry, such as the total station [6] and photogrammetry [7]. However, the photogrammetry has limited measurement range which depends on the camera's field of view, and the total station is not efficient with human collimation.

Recently, the distributed rotary-laser measurement system develops rapidly to accomplish an efficient and accurate three-dimensional coordinate measurement successfully. Take indoor GPS (iGPS) [8] as an example, it has been applied to ensure dimensional accuracy for the shipbuilding in South Korea and it provides an evaluation of the characteristics of ships [9]. Meanwhile, wMPS [10], based on the rotary-laser scanning, has achieved a great success in the large-scale metrology [11],[12]. So, wMPS with the advantages of high-precision and high-efficiency has a broad prospect for the shipbuilding.

The wMPS with multiple transmitters deployed around forms a spatial measurement network to fulfil measurement. The distributed nature makes this measurement network be easily influenced by the environment deeply [13]. For example, the environment of the shipbuilding industry is complex and harsh. Measurement with wMPS is challenged by limited measurement range as well as

IOP Conf. Series: Journal of Physics: Conf. Series **870** (2017) 012009 doi:10.1088/1742-6596/870/1/012009

obstructions of the complex structure which results in large measurement error or even the failure of measurement.

To solve the occlusion problem and ensure the accuracy of spatial measurement network for shipbuilding, this paper introduces a method of reconstructing spatial measurement network by mobile transmitter. The mobile transmitter moves in the measurement area and rapidly involves in the whole measurement network using target points by the spatial resection method. These target points are calibrated by the laser tracker beforehand. When finishing the measurement, this mobile transmitter can be deployed to other places. With the new transmitters joining in, the spatial measurement network is reconstructed and broadened without loss of accuracy. What's more, the occlusion problem is solved with the reconstructed spatial measurement network in the shipbuilding factory.

In this paper, section 2 points out some measurement problems of wMPS faced in the measurement task for the shipbuilding. Section 3 gives a mathematical solution about how to reconstruct the measurement network. In section 4, based on the existing platform, experiences are designed to prove the accuracy and feasibility of this reconstructed measurement network. Finally in section 5, the conclusions of this proposed method are given and the improvement in the future is presented.

2. Description of reconstructing the spatial network for shipbuilding

These is a sketch map below to better figure out the environment in the shipbuilding industry. As the figure 1 shows, it is a part of the hull block assembly in the ship-building factory. The area is a rectangle about 12m*24m. The workpiece is very large and the aisles are just several meters wide. To achieve measurement, the wMPS is used. Four static transmitters are fixed in the corners of this area forming a measurement network. This measurement task has some problems:

1. There are some regions, shown in the figure 1 as "invisible area", which can be seen by one fixed transmitter. So, the measurement can't proceed.
2. Because of the diversity of the workpiece and the restriction by obstructions, no more than two transmitters can be seen at any place.
3. There are also some regions where the intersection angle of transmitters' measuring signal is so large that the measurement condition is bad, like the part reading "large intersection angle "in the figure 1.

Above all, this measurement network can't achieve the measurement either with high accuracy or with overall coverage.

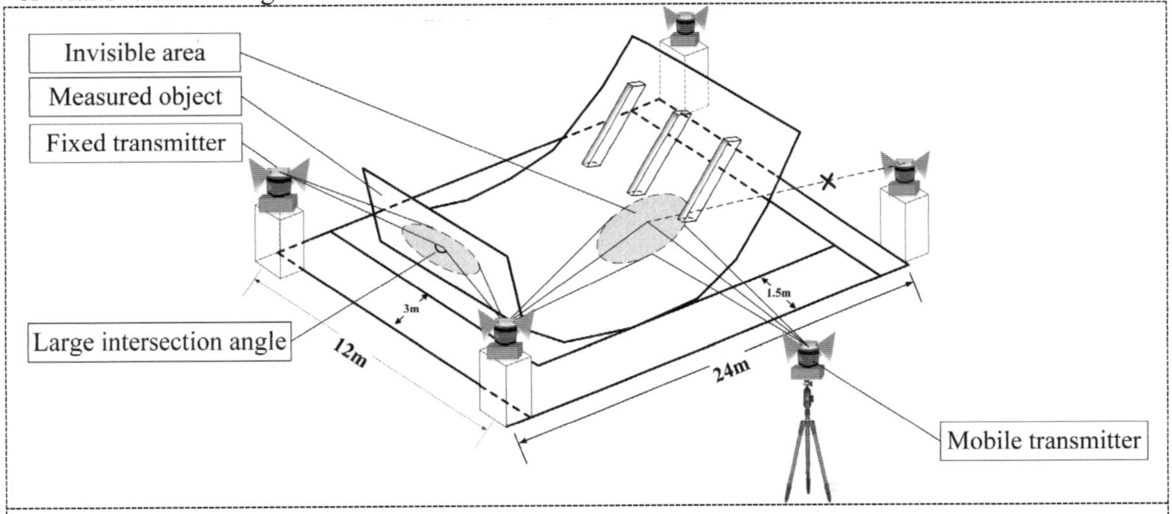

Figure 1. Description of the application with wMPS in the hull block assembly.

In order to deal with these problems, the mobile transmitter is introduced into this system to reconstruct the spatial measurement network shown in figure 1. This mobile transmitter can add into the system quickly without recalibration with the scale bar and it can move to anywhere.

Firstly, in the region where two transmitters can't be seen simultaneously, the mobile transmitter fulfils the measurement tasks with fixed transmitter. It can solve the problem of obstruction efficiently without loss of accuracy.

Secondly, the mobile transmitter can move and fulfil the measurement tasks with other two static transmitters. So, there are three transmitters measuring the coordinate together. The rotary-laser distributed system with three transmitters working together is the best compromise between accuracy and cost [8].

Thirdly, the layout of transmitters is always restricted by the obstruction and the size of the factory. In the large-scale factory, transmitters are often put far away from each other. So, the intersection angle of two static transmitters is random and often less than satisfactory. The mobile transmitter can be deployed to the measurement area in order to ensure that the intersection angle is at the optimum range.

Above all, the new method of reconstructing spatial measurement network is necessary.

3. Mathematical solution

There are two steps to reconstruct the spatial measurement network which are expounded below.

3.1 Establishment of high-precision coordinate control network

In the measurement area, some target points are fixed forming a high-precision coordinate control network, for example putting target points above the workpiece. Also, the target points need to be seen by most or even all of the static transmitters. The distance between coordinate control network and transmitter is less than the transmitter's working distance.

To fulfil the spatial resection method, the target points need to be more than six points. And it's necessary using laser tracker which is more accurate than wMPS to measure the coordinate of target points in the coordinate control network. According to the error analysis [14] of the laser tracker, the range accuracy is better than the angle measurement accuracy. So, based on the length constraint of laser tracker and bundle adjustment [15], a precise coordinate control network is established.

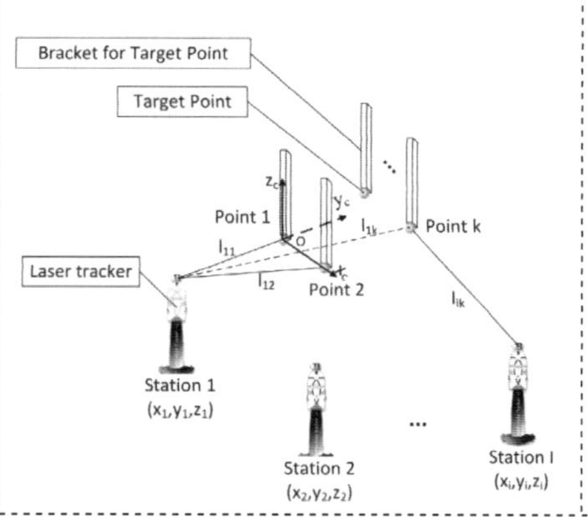

Figure 2. Establishment of the coordinate control network.

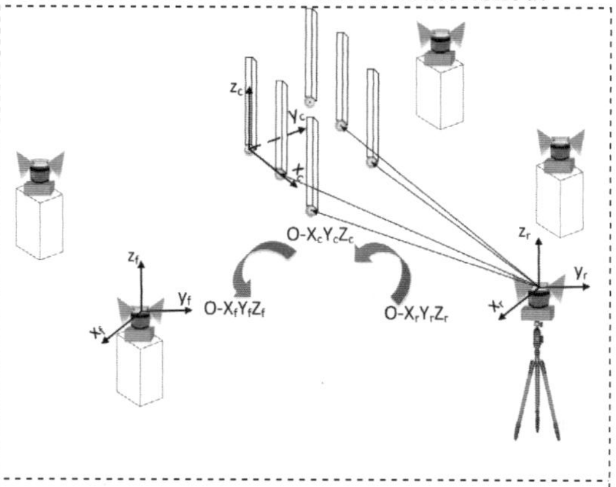

Figure 3. Add the mobile transmitter into the wMPS

As shown in the figure 2, the laser tracker is put at different stations measuring all the target points respectively. The coordinates of target points are (x_{ik}, y_{ik}, z_{ik}) respectively, where i is the position number of laser tracker and k is the number of target points. It defines $O-x_c y_c z_c$ as the coordinate system of the control network measured by the laser tracker when it is put in the first position. With the common target points, the transformation relation between different laser tracker positions is obtained. To reduce the error of transformation, a model of rigid body kinematics [16] is used.

With these transformation matrixes and the target points' coordinate in different laser tracker stations, the coordinates of the laser tracker in different positions can be transferred into one coordinate system, namely the (x_i, y_i, z_i) where i is the position number of the laser tracker. There is an equation below which uses length constraint of the laser tracker to optimize the coordinate of target points.

$$l_{1k}^2 = (x_{1k} - x_i)^2 + (y_{1k} - y_i)^2 + (z_{1k} - z_i)^2 \tag{1}$$

where l_{1k} is the distance between target points and the laser tracker in the first position, k is the number of target points.

To solve this optimum equation, a nonlinear least square is introduced. Among many optimization algorithm, Levenberg-Marquardt algorithm (LM algorithm) is widely used [17],[18].With the initial coordinate of each station (x_i^0, y_i^0, z_i^0) and initial coordinates of target points $(x_{1k}^0, y_{1k}^0, z_{1k}^0)$, the coordinates of target points are optimized and the high-precision coordinate control network $O-x_c y_c z_c$ is established with the final optimized coordinate. This coordinate control network is at least one order of magnitude more accurate than the wMPS and the final coordinates of target points are (x_{1k}, y_{1k}, z_{1k}).

3.2 Using the coordinate control network to rebuild the measurement network with spatial resection method

As shown in figure 3, there are three coordinate systems in the measurement system when adding the mobile transmitter into wMPS: the static transmitter's coordinate system: $O-x_f y_f z_f$ which is calibrated beforehand, the high-precision control network coordinate system: $O-x_c y_c z_c$ which is established in the section 3.1 and the mobile transmitter coordinate system which is established in the mobile transmitter: $O-x_r y_r z_r$. The procedure of adding the mobile transmitter to wMPS is to obtain the transformation relation from the mobile transmitter to the wMPS of static transmitters. If the coordinate of coordinate control network is measured by the measurement network of static transmitter, the transformation relation of these two coordinate systems is obtained.

Then, a method is introduced to obtain the transformation relation between the mobile transmitter coordinate system and control network coordinate system by spatial resection method. Put the mobile transmitter to the certain place and place receivers on the target points. According to the 3-D transformation, the transformation relation of coordinate is as followed:

$$\begin{bmatrix} x_r \\ y_r \\ z_r \end{bmatrix} = \begin{bmatrix} r_1 & r_2 & r_3 \\ r_4 & r_5 & r_6 \\ r_7 & r_8 & r_9 \end{bmatrix} \bullet \begin{bmatrix} x_{1k} \\ y_{1k} \\ z_{1k} \end{bmatrix} + \begin{bmatrix} t_1 \\ t_2 \\ t_3 \end{bmatrix} \tag{2}$$

where (x_{1k}, y_{1k}, z_{1k}) is the coordinate of the target point in this control network coordinate system, k is the number of target points, (x_r, y_r, z_r) is the coordinate of target points in mobile transmitter coordinate system, the rotation matrix $R = \begin{bmatrix} r_1 & r_2 & r_3 \\ r_4 & r_5 & r_6 \\ r_7 & r_8 & r_9 \end{bmatrix}$,translation matrix $T = \begin{bmatrix} t_1 \\ t_2 \\ t_3 \end{bmatrix}$.

When the mobile transmitter works, two laser planes of mobile transmitter pass by the receiver. Each equation of laser plane has been calibrated. The coordinate of receiver satisfies the planar equation below:

$$a_{ni}'x_r + b_{ni}'y_r + c_{ni}'z_r + d_{ni}' = 0 \tag{3}$$

where the $[a_{ni}', b_{ni}', c_{ni}', d_{ni}']$ is the parameter matrix of transmitters' planar equations after rotation [10], n is the number of transmitters, i is the planar equations' number of one transmitter, and $[x_r, y_r, z_r]$ is the coordinate of receiver.

According to the equation 2 and 3, the relation can be established as a matrix format:

$$AX = D \tag{4}$$

where $A = \begin{bmatrix} a_{11}'x_{11} & a_{11}'y_{11} & a_{11}'z_{11} & b_{11}'x_{11} & b_{11}'y_{11} & b_{11}'z_{11} & c_{11}'x_{11} & c_{11}'y_{11} & c_{11}'z_{11} & a_{11}' & b_{11}' & c_{11}' \\ a_{12}'x_{11} & a_{12}'y_{11} & a_{12}'z_{11} & b_{12}'x_{11} & b_{12}'y_{11} & b_{12}'z_{11} & c_{12}'x_{11} & c_{12}'y_{11} & c_{12}'z_{11} & a_{12}' & b_{12}' & c_{12}' \\ & & & & & \vdots & & & & & & \\ a_{12}'x_{1k} & a_{12}'y_{1k} & a_{12}'z_{1k} & b_{12}'x_{1k} & b_{12}'y_{1k} & b_{12}'z_{1k} & c_{12}'x_{1k} & c_{12}'y_{1k} & c_{12}'z_{1k} & a_{12}' & b_{12}' & c_{12}' \end{bmatrix}$

and $D = [-d_{11}', \ -d_{12}', \ \cdots -d_{11}', \ -d_{12}']$

In the equation above, the rotation matrix and translation matrix are unknown and its parameter $X = \left[r_1, r_2, r_3, r_4, r_5, r_6, r_7, r_8, r_9, t_1, t_2, t_3\right]^{\mathrm{T}}$ is to be calculated. Because of the error of two plane equations' parameters and the coordinates of control network measured, there is no accurate value of the matrix X by the linear least-square algorithm. Also, the value of intercept d_{11}' or d_{12}' is so small that the matrix D is nearly a zero matrix. So, a preliminary calculation and nonlinear least-square algorithm is used to solve these equations accurately. In the following analysis, the incorrect value of the equation 5 is named as X_r, the initial value of optimization algorithm is named as X_i and the exact solution of equation is named as X_e.

Firstly, the incorrect value X_r of the equation $AX = D$ is calculated by linear solution directly. The matrix D is nearly a zero matrix, so this is a homogeneous equation and there is a basic set of solutions. The initial value X_i of X has a fixed proportion with the basic set of solutions. Then, a compensation of ratio is to be done. According to the property of orthogonality of the rotation matrix, the parameters r_1, r_2, r_3 satisfy the equation below:

$$r_1^2 + r_2^2 + r_3^2 - 1 = 0 \tag{5}$$

So, the ratio can be got by the equation above and the initial value X_i of the optimization algorithm can be calculated by X_r multiplying the ratio.

Secondly, the nonlinear least-square method with Levenberg-Marquardt iterative method is to optimize the solution of the equation 5. The parameters of the rotation and translation matrix $r_1, r_2, r_3, r_4, r_5, r_6, r_7, r_8, r_9$ satisfy the orthogonal constraints.

Finally, considering the strong restriction of orthogonal matrix, the penalty function is used to deal with these constraints. With the equation 4 and equation 7, the objective function can be written as below:

$$E = \sum_{i=1}^{2} F_{ik}^2 + M\sum_{j=1}^{6} f_j^2 = \min \tag{6}$$

where M is the penalty factor.

Above all, the accurate rotation and transition matrix can be calculated accurately and the mobile transmitter adds into the wMPS formed by static transmitters. The spatial measurement network is reconstructed by this mobile transmitter.

4. Experiments

4.1 Experiment platform set-up

To validate the feasibility and accuracy of reconstructed measurement network by the mobile transmitter of wMPS, two contrast experiments are designed to evaluate the performance of reconstructed measurement network. The first one is to validate the accuracy of point-to-point length measurement. The second one is to compare with the laser tracker in the coordinate measurement.

As shown in figure 4, two static transmitters and a mobile transmitter are prepared with its intrinsic parameters calibrated. One transmitter's azimuth angular uncertainty [19] was about 2.4″ (with the confidence probability of 99.73%). Two transmitters have formed a measurement network and a mobile transmitter is prepared to add into the network. Then, a new measurement network with only a static transmitter and a mobile transmitter forms, and it is called the new measurement network in the following passage. These experiments are to verify the feasibility and accuracy of this new measurement network. The coordinate control network is fixed on the roof and calibrated by the Leica AT901—LR laser tracker. About this laser tracker, the maximum permissible error (MPE) of the angular accuracy is ±15μm+6μm/m, and the MPE of interferometer accuracy is ±0.4μm+0.3 μm/m. The receiver of wMPS has the same size with the spherically mounted retroreflector (SMR) of the laser tracker in order to make a comparison.

Figure 4. Platform of the proposed method

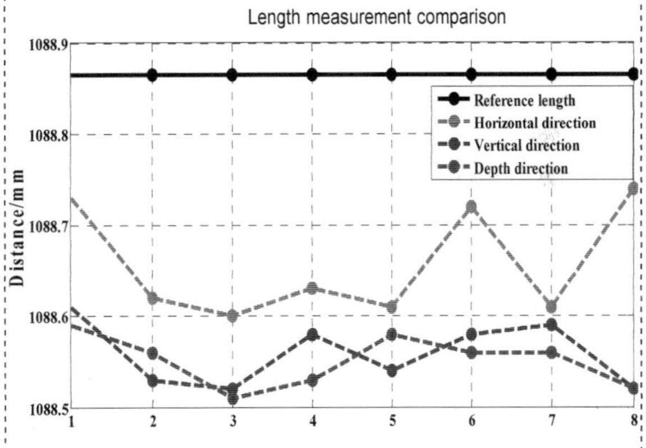

Figure 5. Point-to-point length measurement

4.2 Verification of point-to-point length measurement

According to the Standards "ASME B89.4.19" [20] on dimensional metrology, the experiment of point-to-point length measurement test is designed. This experiment is to compare the point-to-point length measured by the new measurement network with a scale bar. This scale bar is calibrated by the laser tracker as the reference length. On both sides of the scale bar, there are spherical housings to place either a receiver or a SMR. When comparing the length, the scale bar is put about three kinds of attitudes, in the horizontal direction, in the vertical direction and in the depth direction. To improve the accuracy of this experiment, it performs ten times at each position, and has eliminated the gross error. The data is shown as figure 5:

As shown from the figure above, comparing with the laser tracker, the point-to-point length measurement error of the new measurement network in three directions is less than 0.4mm. The length error in the horizontal direction is less than the error of other directions. The accuracy of length measurement in each direction depends on either the accuracy of wMPS or the layout of coordinate control network, and the coordinate control network provides the strong constraints in space to ensure the accuracy of wMPS. In the figure 5, the length measured by the new measurement network is always less than the one measured by the laser tracker. It's mainly because the receiver's center of sphere is not at the same position with SMR.

4.3 Verification of coordinate measurement

The new measurement network is used for coordinate measurement, so the accuracy of coordinate measurement is necessary to ensure. In this experiment, the coordinate system of the proposed method is the same with the laser tracker coordinate system by coordinate transformation of some non-collinear common points. There are nine groups of coordinate to be compared in the space and each coordinate measures ten times at one place. The comparison results are listed in table 1.

Table 1. Coordinate measurement experiment

number	Coordinate measured by the laser tracker (mm)			Coordinate measured by the proposed method(mm)			Deviation[a] (mm)
	x_c	y_c	z_c	x_t	y_t	z_t	Δ
1	-3677.622	2991.999	1031.473	-3677.62	2992.13	1031.34	0.215
2	-3667.079	3009.469	230.582	-3667.24	3009.34	230.33	0.326
3	-3661.098	3018.214	-570.162	-3661.10	3017.69	-570.21	0.527
4	2263.418	5616.739	-511.049	2263.00	5617.12	-510.85	0.599
5	2258.012	5605.601	297.892	2257.65	5605.91	298.07	0.508
6	2260.388	5593.652	1094.217	2260.49	5593.53	1094.07	0.217
7	-1251.391	6745.954	1645.435	-1251.49	6745.74	1645.39	0.240
8	-2125.118	6383.763	1632.785	-2125.27	6383.68	1632.71	0.189
9	-877.979	5752.818	-247.201	-878.38	5752.85	-247.28	0.410

[a] The deviation: $\Delta = \sqrt{(x_t - x_c)^2 + (y_t - y_c)^2 + (z_t - z_c)^2}$

As shown in table 1, the absolute position error of spatial coordinate measurement comparing with the laser tracker is less than 0.6mm. In fact, because of the instrumental error of the laser tracker and wMPS, the coordinate transformation method with common points inevitably induces some system error in the comparison experiment. Even though the error exists, this experiment still verifies the accuracy and feasibility of the proposed method.

5. Conclusions

A method of reconstructing spatial measurement network by mobile rotary-laser transmitter for shipbuilding is presented. This mobile transmitter can be placed flexibly in the measurement area so that it can adapt to various complex environments for shipbuilding and ensure measurement accuracy. With more than six target points, the mobile transmitter is able to add into the measurement network with the spatial resection method quickly and reconstruct the measurement network. Also, the establishment of precise coordinate control network with several target points is used to ensure the measurement accuracy of the optimized measurement network.

To prove the accuracy and feasibility of this reconstructed measurement network with mobile transmitter, two experiments are designed at a distance of five meters away from the transmitter. Comparing with the laser tracker, the result shows that the point-to-point length measurement accuracy in three directions is better than 0.4mm and the absolute positioning error is less than 0.6mm. Therefore, the spatial measurement network reconstruction method by mobile rotary-laser transmitter for shipbuilding is considered to be a feasible and efficient solution in large-scale complex environment.

With regard to this measurement network reconstruction for the complex environment by a mobile transmitter, more detailed work is to be improved in the future, such as combining inclinometer with the transmitter to restrict transmitter's degree of freedom. In addition, to improve the accuracy of the measurement network, the combined adjustment is needed.

References

[1]. Estler, W. T., Edmundson, K. L., Peggs, G. N., & Parker, D. H. 2002 Large-scale metrology–an update. *CIRP Annals-Manufacturing Technology*, 51(2), 587-609..
[2]. Guan-yun, L. I. 2001 The State of the Art and Applications of the Industrial Measuring Systems.

Engineering of Surveying and Mapping, 2, 008.

[3]. Peggs, G. N., Maropoulos, P. G., Hughes, E. B., Forbes, A. B., Robson, S., Ziebart, M., & *Muralikrishnan*, B 2009 Recent developments in large-scale dimensional metrology. *Proceedings of the Institution of Mechanical Engineers, Part B: Journal of Engineering Manufacture*, 223(6), 571-595.

[4]. Peggs, G. N., Maropoulos, P. G., Hughes, E. B., Forbes, A. B., Robson, S., Ziebart, M., & Muralikrishnan, B. 2009 Portable Metrology and Dimensional Control in Integrated Shipbuilding Assembly Process. *Proceedings of the Institution of Mechanical Engineers, Part B: Journal of Engineering Manufacture*, 223(6), 571-595.

[5]. Huang F, Li W Y 2015 The application of total station hull block assembly. *Marine Technology*.

[6]. Kyle, S. A., Robson, S., Chapman, D. P., Cross, P. A., Iliffe, J. C., & Oldfield, S. 2001 Understanding Large Scale Metrology. *Understanding Large Scale Metrology*.

[7]. Koelman, H. J. 2010 Application of a photogrammetry-based system to measure and re-engineer ship hulls *and* ship parts: An industrial practices-based report. *Computer-Aided Design*, 42(8), 731-743.

[8]. Maisano, D. A., Jamshidi, J., Franceschini, F., Maropoulos, P. G., Mastrogiacomo, L., Mileham, A., & *Owen*, G 2008 Indoor GPS: system functionality and initial performance evaluation. *International Journal of Manufacturing Research*, 3(3), 335-349.

[9]. Park, J. T., Hayden, D. D., Kim, J. H., & Melendez, M. P. 2008 A New Methodology for the *Evaluation* of the Maneuvering Characteristics of Surface Ship Models. *In ASME 2008 Fluids Engineering Division Summer Meeting collocated with the Heat Transfer, Energy Sustainability, and 3rd Energy Nanotechnology Conferences* (pp. 393-398). American Society of Mechanical Engineers.

[10]. Zhao, Z., Zhu, J., Lin, J., Yang, L., Xue, B., & Xiong, Z. 2014 Transmitter parameter calibration of the workspace measurement and positioning system by using precise three-dimensional coordinate control network. *Optical Engineering*, 53(8), 084108-084108

[11]. Xue, B., Zhu, J., Zhao, Z., Wu, J., Liu, Z., & Wang, Q. 2014 Validation and mathematical model of *workspace* Measuring and Positioning System as an integrated metrology system for improving industrial robot positioning. *Proceedings of the Institution of Mechanical Engineers, Part B: Journal of Engineering Manufacture*, 228(3), 422-440.

[12]. Xue, B., Zhu, J., Yang, L., Zhao, Z., & Ye, S. 2014 The Application of the wMPS in Airplane Level Measurement. *Photoelectric engineering*, 41(8), 22-26.

[13]. Galetto, M., & Pralio, B. 2010 Optimal sensor positioning for large scale metrology applications. *Precision Engineering*, 34(3), 563-577.

[14]. JIN, Z. J., LI, J. X., YU, C. J., & KE, Y. L. 2015 Registration error analysis and evaluation in large-volume metrology system. *Journal of Zhejiang University (Engineering Science)*, 4, 007.

[15]. *Predmore*, C. R. 2010 Bundle adjustment of multi-position measurements using the Mahalanobis distance. *Precision engineering*, 34(1), 113-123.

[16]. Geng, N., Zhu, J., Lao, D., & Ye, S. 2010 Theory and algorithm of coordinate system registration based on rigid body kinematics. *Chinese Journal of Sensors and Actuators*, 23(8), 1088-1092.

[17]. *Ranganathan*, A. 2004 The levenberg-marquardt algorithm. *Tutoral on LM algorithm*, 1-5.

[18]. Lourakis, M. I. 2005 A brief description of the Levenberg-Marquardt algorithm implemented by *levmar*. *Foundation of Research and Technology*, 4(1).

[19]. XIONG, Z., ZHU, J., GENG, L., REN, Y., YANG, X., & YE, S. 2012 Verification of angle measuring uncertainty for workspace Measuring and Positioning System. *Chinese Journal of Sensors and Actuators*, 2, 019.

[20]. American National Standards Institute. 2006 *Performance Evaluation of Laser-based Spherical Coordinate Measurement Systems*. American Society of Mechanical Engineers.

Acknowledgments
This work was funded by the National Natural Science Foundation of China (Grant No. 51405338, 51475329). The authors wish to express their sincere appreciation to editors and comments from the viewers would be appreciated very much.

IOP Conf. Series: Journal of Physics: Conf. Series **870** (2017) 012010 doi:10.1088/1742-6596/870/1/012010

DMI measurements impact on a position estimation with lack of GNSS signals during Mobile Mapping

K Bobkowka[1],G Nykiel[1], P Tysiąc[1]

[1]Department of Geodesy, Faculty of Civil and Environmental Engineering, Gdansk University of Technology, Gabriela Narutowicza str. 11/12, 80-233 Gdansk, Poland

pawel.tysiac@pg.gda.pl

Abstract. Nowadays, Mobile Laser Scanning is common in use in addition to geodesy measurements. The data which are provided by the system characterizes with high precision and flexibility. To precise mapping, the accuracy of the data should be maintained. In Poland, according to the minister's dispositions, the accuracy of the data should not exceeded 10 cm. With fully operated system it is easy to uphold, but there is a situation when a signal from an INS is not enough to preserve it. This paper is presenting the solution of a DMI use in Mobile Laser Scanning measurements as the support for position estimation during lack of satellites signal situation when the vehicle with the platform was entered the tunnel. To comparison the results a several of entrances was performed. This research helps understand the use of DMI in mobile data acquisition in different acquiring situations.

1. Introduction

Mobile Laser Scanning data provides data which characterizes with high accuracy, precision and point density. Nowadays it is very important to gather the spatial information very fast to minimalize the cost of the measurements which could be noticed in traditional geodesy measurements methods. It is very important to maintain the continuously of the data accuracy, because of the project's assumptions where the accuracy was estimated [1]. Sometimes it is very problematic, considering the field shutters like high buildings, trees or the change of DOP (dilution of precision) factor [2], [3], [4]. For this study, the lack of satellites was taken in consideration in case of measurements in a tunnel. To maintain the position estimation, the GNSS/INS was used with DMI (distance measurement instrument) device as an additional equipment. This device measures wheel revolutions of the vehicle. The characteristic of obtaining data slows the growth of the inertial error which influence on the quality of the data. We present comparison of measurements in obstructed area, when DMI was in use and when DMI was excluded. The measurements were performed during laser scanning scanned object which was a tunnel under the Martwa Wisla in Gdansk. The use of laser scanning for this type of building is quite wide. This measurement method can be used, for example, for rockmass classification [5] or for tunneling monitoring [6]. It is also used to detect cracking of concrete objects [7], [8], [9], [10].

2. Experiment realization method

The data was collected by using Riegl VMZ-400 system, provided by Apeks Company based in Gdansk, Poland. Before the experiment realization measurements, the proper calibration system should be performed. It consists of measuring the offsets between the phase center of the device and DMI,

GNSS/IMU [11] and proper alignment of the angles roll, pitch and yaw. To proper align these angles, it is required to perform the static and dynamic alignment, where the static alignment refers to 5 minutes spatial data gathering through GNSS receiver and dynamic to accelerate and to decelerate the vehicle where the system is placed. That processes are crucial in addition to have proper alignment of points with expected accuracy. For the realization of the experiment, a several passing through the tunnel was performed. The tunnel is 1400 meters long. In the beginning of the measurements, the availability of the satellites was good and after entering the tunnel signal was lost. The trajectory was aligned in two different ways in post-processing mode. First, when the system was fully operational (GNSS/IMU + DMI) and the second, when the trajectory was aligned without DMI. It helps the authors to understand the phenomena of DMI impact on the point cloud records accuracy and precision and achieve the main goal of this paper which is showing the impact of the DMI measurements on mobile mapping.

3. Positioning results

The main goal of this paper is to show the crucial role of using the DMI device when the GNSS signal is lost. In Figure 1 standard deviations (STD) of position and angles derived from GNSS/IMU measuring system are presented. During the measurements time 7 passes were made through the tunnel. This is clearly seen in the figure, because position STD values increased. The highest values were obtained for North and East component of the position. However, it is worth to notice, that when the DMI device was used STDs decreased by about 25%, 45% and 25% for North, East and Up component respectively. In Fig. 1 STD of angles derived from IMU device are also presented. It can be seen that usage of DMI has not significant improvement in case of STD of roll and pitch angles. Values obtained with and without DMI are very similar. Interesting results are visible in case of heading angle. It is seen that usage DMI measurements decreases STD values by approximately 10%. However, what is seen in the figure, improvements occur throughout the duration of the measurements, not only when the tunnel was scanned.

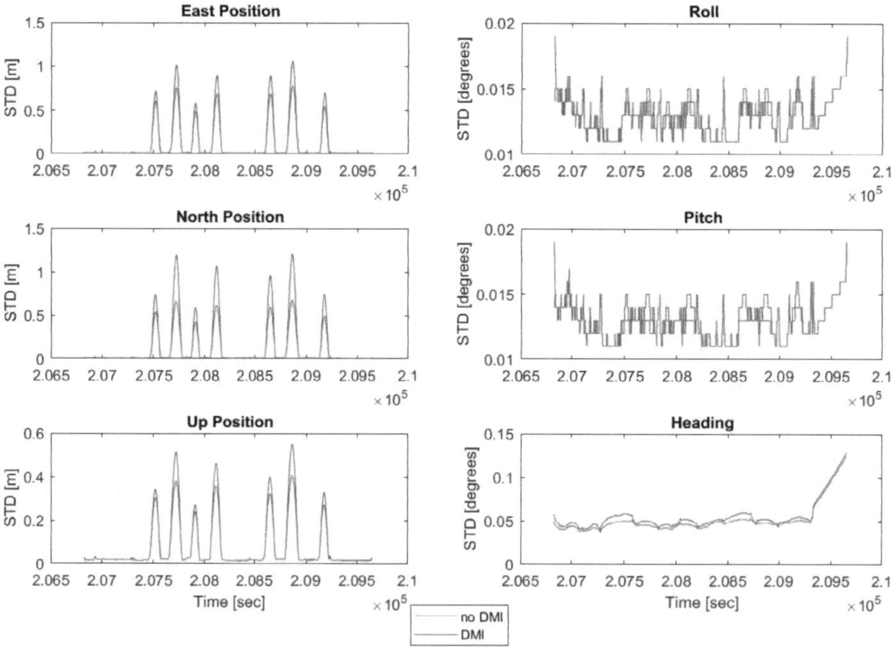

Figure 1. Standard deviation values for position (left) and angles (right) measurements derived from GNSS/IMU system (orange line: with DMI, blue line: without DMI).

In Table 1 and Table 2 numerical values for sample passage through a tunnel are presented. The first record refers to situation when the platform was on the open space where satellites geometry was good and DOP factor had low value. There is no significant differences between measuring the system with or without the DMI. The second one is presenting entering the tunnel where availability of the satellites were constrained because the appearance of wall shutters, housing the tunnel. The third and the forth are presenting STD of a position estimation in a half way of a tunnel (the third) and at the end (the forth). As is seen the values are significantly high in order to maintain the accuracy and precision up to 10 cm. However, as we mentioned above, usage of DMI device together with GNSS/IMU. In Table 2 it is also seen impact of DMI of the heading angle measurements, which does not matter whether the measurement took place, in or before the tunnel. Thus, we conclude that be an effect of the distance measurement indicator influence on the heading angle value.

Table 1. Comparison of position (in topocentric coordinates) standard deviations (in meters) between solution when DMI was used and was not.

No.	GPS time	East STD DMI	East STD no DMI	North STD DMI	North STD no DMI	Height STD DMI	Height STD no DMI
1	**207230.52026**	0.014	0.009	0.019	0.014	0.012	0.019
2	**207561.56578**	0.080	0.052	0.093	0.048	0.092	0.073
3	**207694.94101**	0.495	0.636	0.493	0.822	0.313	0.412
4	**207722.55106**	0.752	1.008	0.653	1.190	0.379	0.509

Table 2. Comparison of Roll, Pitch and Heading standard deviations (in degrees) between solution when DMI was used and was not.

No.	GPS time	Roll STD DMI	Roll STD no DMI	Pitch STD DMI	Pitch STD no DMI	Heading STD DMI	Heading STD no DMI
1	**207230.52026**	0.012	0.012	0.012	0.012	0.042	0.046
2	**207561.56578**	0.013	0.013	0.013	0.013	0.050	0.056
3	**207694.94101**	0.013	0.014	0.013	0.014	0.045	0.050
4	**207722.55106**	0.013	0.015	0.013	0.015	0.045	0.050

4. Mobile mapping results

Presented positioning results should have impact on mobile laser scanning. In order to verify this, we consider point cloud quality obtained during measurement with and without DMI. First, the situation where the DMI device on the open space with good DOP was taken on consideration. The section of the tunnel before enter is presenting as a point cloud in Figure 2.

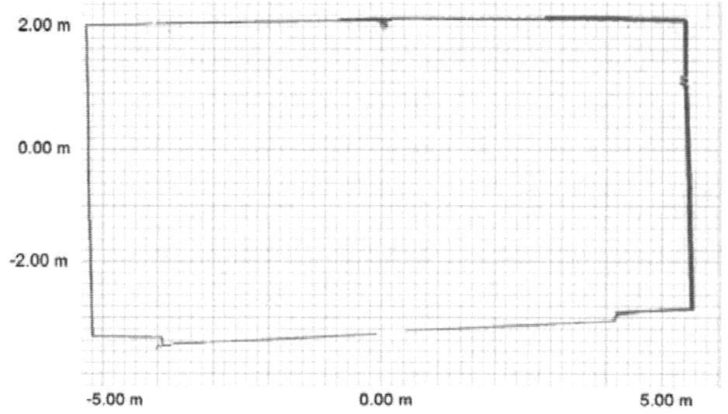

Figure 2. The section of the tunnel when the system was fully operational.

As its seen on the figures, the data was align correctly, based on the structure of the point cloud. The position error growths when the vehicle with the platform entered a tunnel. The authors present the results of the experiment when the GNSS signal was lost for more than 2 minutes. Figure 3 shows the result of fully operational mobile mapping system after losing the GNSS signals for more than 2 minutes, which (based on speed limits in Poland) equals 1.4 km. The difference between three records is up to 40 cm, but it has to be mentioned that two of them are aligned with the error, not exceeded 10 cm. In Figure 4, the point cloud presenting the results of measurements when the DMI was not used is shown. This solution is very important in case of placing the mobile platform on a vessel [12] and it is crucial to understand the phenomena of the measurements in order to increase the accuracy with no DMI solution in mobile mapping.

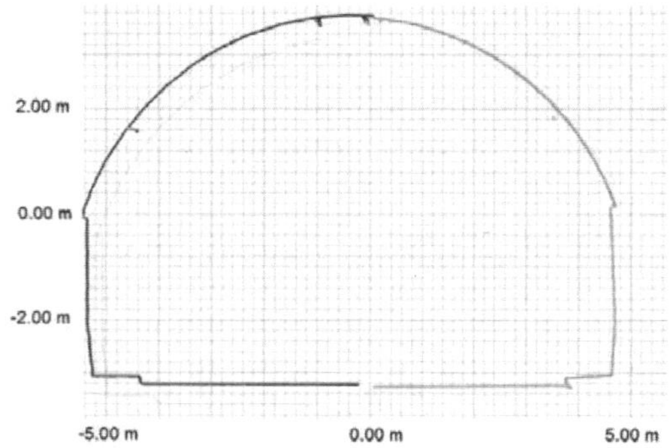

Figure 3. The three point cloud records, showing the align errors fully operational systems (GNSS/IMU + DMI).

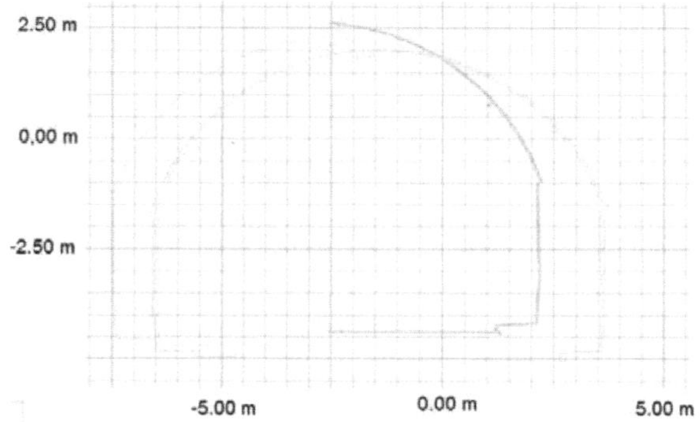

Figure 4. The three point cloud records, showing the align errors of the system without DMI.

In Figure 5 trajectories derived from the mobile scanning platform are shown. By red color we marked trajectories obtained with GNSS/IMU + DMI system, and by green color when DMI was excluded. On left side of the figure, three trajectories measured during passage the tunnel are presented. The differences between solution with and without DMI are clearly visible and reach even a few meters. This is caused by positioning incompatibilities, which we described. As it is seen, outside the tunnel the differences between trajectories can be neglected.

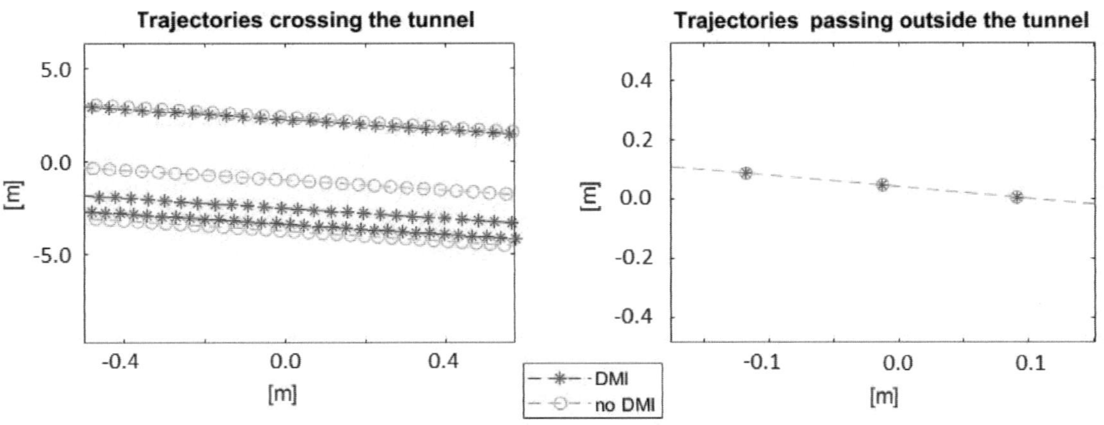

Figure 5. Trajectories derived from the mobile scanning platform during crossing the tunnel (left) and outside the tunnel (right).

5. Conclusions

Mobile mapping is still new spatial data acquisition method and have a lot of constrains, which allows to develop new measuring solutions which could be successfully used in geodesy and cartography or even civil engineering. We presented impact of DMI measurements on the mobile mapping in reference to lack of the GNSS signals. We proved, that a distance measurement indicator device is crucial in use when the GNSS signals is lost. We obtained that DMI can increase the precision of position by about 25% in obstructed area. Also, some improvements of heading angle derived from IMU were observed. Usage of DMI decreased its STD by about 10% during throughout the measurement period.

To check how the obtained positioning results impact on the mobile mapping, we presented comparison of the point clouds and trajectories. We can stated that differences of trajectories between measurements with DMI and without DMI can reached up to few meters during passage the tunnel. However, even the results of the trajectory STD were up to 1 meter with the use of the DMI, the precision of the point cloud does not exceed 40 cm in reference to three records, two of them, are align with an error which does not exceed 7 cm.

It has to be mentioned that mobile systems could be used when there is no possibility to mount the DMI (for example mobile platform, placed on a vessel [13]). In that case, a geometry of satellites is good, when a vessel ship on an open space, but when organize a measurements on the rivers, in a high urbanized area, the situation could be different [14], [15].

For the further research, the authors are going to improve the quality of the data, when the DMI could be used, but there will be lack of the satellites situation and when the DMI could not be used.

References

[1] Ogle J, Guensler R, Bachman W, Koustak M, and Wolf J 2002 Accuracy of global positioning system for determining driver performance parameters *Transp. Res. Rec. J. Transp. Res. Board* **1818** pp. 12–24.

[2] Jing H, Slatcher N, Meng X and Hunter G 2016 Monitoring capabilities of a mobile mapping system based on navigation qualities *Int. Arch. Photogramm. Remote Sens. Spat. Inf. Sci.*, 12-19 July 2016, Prague, Czech Republic

[3] Nowak A 2016 Dynamic GNSS Mission Planning Using DTM for Precise Navigation of Autonomous Vehicles *J. Navig.*

[4] Nowak A 2015 The Proposal to 'Snapshot' RAIM method for GNSS vessel receivers working in poor space segment geometry *Polish Marit. Res.* **22** no. **4(88)**, pp. 3–8

[5] Fekete S, Diederichs M, and Lato M 2010 Geotechnical and operational applications for 3-

dimensional laser scanning in drill and blast tunnels *Tunn. Undergr. Sp. Technol.*, **25** no. **5**, pp. 614–628

[6] Yoon J S, Sagong M, Lee J S and Lee K S 2009 Feature extraction of a concrete tunnel liner from 3D laser scanning data *NDT E Int.* **42** no. **2** pp. 97–105

[7] Janowski A Nagrodzka-Godycka K Szulwic J and Ziolkowski P 2016 Remote sensing and photogrammetry techniques in diagnostics of concrete structures *Comput. Concr.* pp. 335–342,

[8] Janowski A, Jurkowska A, Przyborski M, Sobieraj A, Szulwic J, Wróblewska D and Wieczorek B 2014 Improving the quality of education through the implementation of the diplomas and group projects during engineering studies in cooperation with employers *EDULEARN14 Proceedings (2014)* pp. 1837–1843.

[9] Nagrodzka-Godycka K Szulwic J and Ziółkowski P 2015 Method of selective fading as a educational tool to study the behaviour of prestressed concrete elements under excess loading *8th International Conference of Education, Research and Innovation, ICERI2015 Proceedings*, 2015 no. **ISBN: 978-84-608-2657-6** pp. 472–479.

[10] Nagrodzka-Godycka K, Szulwic J and Ziółkowski P 2016 Accuracy improvement of the prestressed concrete structures precise geometry assessment by use of bubble micro-sampling algorithm *16th International Multidisciplinary Scientific GeoConference SGEM 2016* DOI: 10.5593/SGEM2016/B22/S10.102 pp. 799-806,

[11] Chen C, Liu H, Liu Y and Zhuo X 2013 High accuracy calibration for vehicle-based laser scanning and urban panoramic imaging and surveying system *Proc. SPIE 8917, MIPPR 2013 Multispectral Image Acquis. Process. Anal. 89170Y* doi:10.1117/12.2031466

[12] Burdziakowski P, Janowski A, Kholodkov A, Matysik K, Matysik M, Przyborski M, Szulwic J, Tysiac P and Wojtowicz A 2015 Maritime laser scanning as the source for spatial data *Polish Marit. Res.* **22** no. **4** pp. 9–14, 2015.

[13] Shi B, Lu X, Yang F, Zhang C, Lv Y and Min C 2017 Shipborne Over- and Under-Water Integrated Mobile Mapping System and Its Seamless Integration of Point Clouds *Mar. Geod.*, pp. 1–19

[14] Guan H, Li J, Yu Y, and Liu Y 2016 Geometric validation of a mobile laser scanning system for urban applications *2015 ISPRS International Conference on Computer Vision in Remote Sensing. International Society for Optics and Photonics* pp. 990108–990108.

[15] Kwoczynska B, Sagan W and Dziura K 2016 Elaboration and Modeling of the Railway Infrastructure Using Data from Airborne and Mobile Laser Scanning *2016 Baltic Geodetic Congress (BGC Geomatics)* pp. 106–115.

IOP Conf. Series: Journal of Physics: Conf. Series **870** (2017) 012011 doi:10.1088/1742-6596/870/1/012011

WAMS measurements pre-processing for detecting low-frequency oscillations in power systems

P Y Kovalenko

Automated Electrical Systems Department, Ural Federal University named after the first President of Russia B.N. Yeltsin. 19 Mira st., Yekaterinburg, Russia

pkovalenko@urfu.ru

Abstract. Processing the data received from measurement systems implies the situation when one or more registered values stand apart from the sample collection. These values are referred to as "outliers". The processing results may be influenced significantly by the presence of those in the data sample under consideration. In order to ensure the accuracy of low-frequency oscillations detection in power systems the corresponding algorithm has been developed for the outliers detection and elimination. The algorithm is based on the concept of the irregular component of measurement signal. This component comprises measurement errors and is assumed to be Gauss-distributed random. The median filtering is employed to detect the values lying outside the range of the normally distributed measurement error on the basis of a 3σ criterion. The algorithm has been validated involving simulated signals and WAMS data as well.

1. Introduction

It is common knowledge that the essential source of information on the power system equipment current operation state is the combination of the equipment performance parameters measurements. At the same time measurements validity i.e. degree of belief is defined by these measurements accuracy [1, 2]. To date large-scale power systems impose strict requirements on measurement systems accuracy for the purposes of monitoring and improving stability, controllability and security [3–5]. Wide-area monitoring systems (WAMS) based on phasor measurements technology have been implemented in many power systems in order to meet these requirements [6].

Nevertheless there are errors inherent in WAMS data inevitably. These errors may be due to various causes including measurement errors of instrument transformers, secondary circuits and Phasor measurement units (PMUs), data transfer failures etc. [7]. These reasons may lead to emerging outliers [8] i.e. values standing apart from the sample collection. While it is unlikely that instrument transformers and secondary circuits measurement errors would change rapidly over time it is still typical for PMU themselves to cause outliers during transients which is proved by the corresponding research [9–11].

This particular feature influences significantly the low-frequency oscillations (LFO) express-analysis (detecting and estimating parameters) performance capability [12]. It is of crucial importance to pinpoint outliers in WAMS data received online and correct them without making unnecessary and undesired changes to surrounding data.

IOP Conf. Series: Journal of Physics: Conf. Series **870** (2017) 012011 doi:10.1088/1742-6596/870/1/012011

2. Outliers elimination algorithm

The corresponding algorithm has been developed for the purpose of locating the outliers. It is necessary to define the maximum duration or number of data points considered to be the outlier. The outlier duration is subject to WAMS data sampling rate. For example in case the sampling rate is 50 Hz i.e. 1 measurement point per every 0.02 s and LFO of 5 Hz frequency arise the oscillations half-period is 5 points long. As the algorithm is aimed at preserving the oscillations data it is recommended to set the outlier maximum duration equal to 4 points. Nevertheless this parameter could be adjusted depending upon the field of application.

The median filter has been selected to be utilized within the algorithm due to its capability to leave the data points of the signal being processed unaltered in case the outlier duration does not exceed the half of the window width. This is a significant advantage compared to smoothing filtering, which alters all data points.

First, median filter is applied to the data $x(i)$ under consideration with filter window defined as $2n+1$ where n is the predefined maximum outlier duration:

$$x(i) \rightarrow m(i) . \tag{1}$$

After that the filtering result $m(i)$ is subtracted from the initial data:

$$d(i) = x(i) - m(i) , \tag{2}$$

and standard deviation σ_m is calculated for the obtained difference:

$$\sigma_m = \sqrt{\frac{1}{k-1} \sum_{j=n-k}^{n-1} \left(d_i - \overline{d}\right)^2} , \tag{3}$$

where $\overline{d}$ is the expected value of $d(i)$ within the $[n-k; n-1]$ interval.

According to equation (3), the standard deviation σ_m is calculated not for the whole sample collection but at every sequential point of it. This allows for time variations of non-regular component of the signal.

Then all points where absolute $d(i)$ values exceed $\overline{d} + 3\sigma_m$ are considered to be outliers and replaced by linear sections. This approach is reasonable when equipment measurement errors distribution is close to Gaussian which is correct for currently employed PMUs [4].

3. Algorithm experimental validation

The algorithm is validated using MATLAB software package. The validation involves simulated signals and actual WAMS data.

3.1. Validation using simulated data

In order to prove the algorithm effectiveness it has been validated using complex harmonic noise-contaminated signal with artificially added outliers:

$$0.3\sin(0.05t) + \sin(1.3 \times 0.05t) + 0.9\sin(4.2 \times 0.05t) + 0.05 \times randn , \tag{4}$$

where $randn$ is Gaussian random variable with zero mean.

The outlier maximum duration for the algorithm is set to 4 points while the last added outlier duration is intentionally made to be 5 points. Figures 1–4 present the steps of the algorithm.

Figures 5–6 present the interval of the initial data before and after outliers elimination. The last outlier duration of which was set to exceed the maximum eliminated outlier duration is preserved in the data.

Since the algorithm has been designed to eliminate random measurement errors maximum outlier duration of more than 4 points should not be set unless it is reliably known that the data in question contains long-duration outliers. Otherwise data without outliers might be undesirably modified.

Validation using simulated data has shown that the algorithm is capable of eliminating outliers of predefined duration with no additional unwanted changes made to the signal on the intervals without outliers.

IOP Conf. Series: Journal of Physics: Conf. Series **870** (2017) 012011 doi:10.1088/1742-6596/870/1/012011

3.2. Validation using actual WAMS data

It was outlined that outliers are inherent for the PMU data during transients. Thus disturbance WAMS data from PMUs installed on power stations in Russia was involved in algorithm validation. Figures 7–8 show the frequency recorded during a disturbance and outliers elimination result for this signal.

The outlier is clearly visible in figure 7 immediately after the disturbance. This outlier implies difficulties while identifying the low-frequency oscillations following after the disturbance. Automatic outlier elimination enables avoiding this effect.

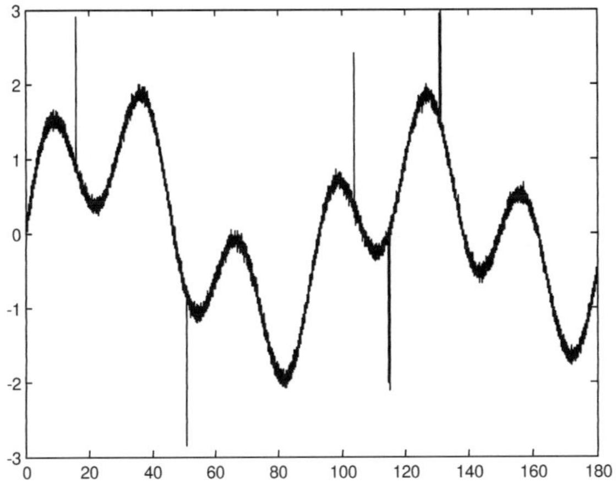

Figure 1. Simulated data.

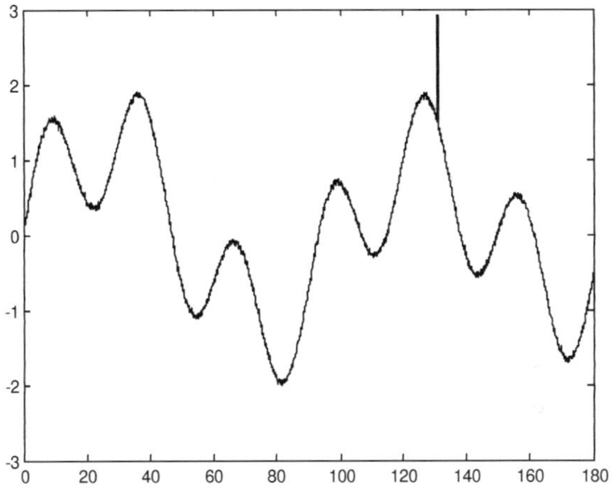

Figure 2. Filtering result.

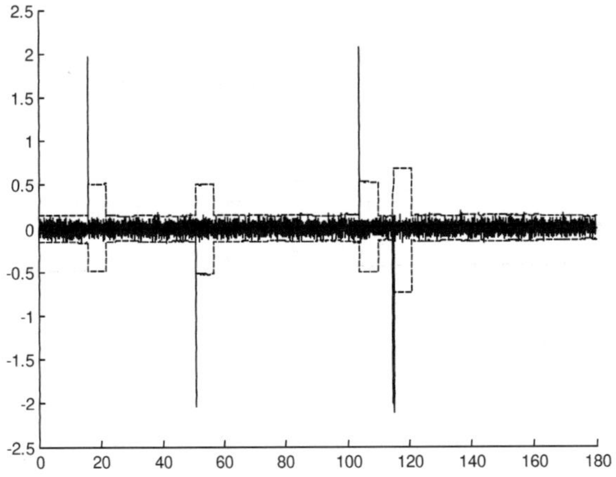

Figure 3. Difference between the initial data and the filtering result; $\overline{d} + 3\sigma_m$ (dash line).

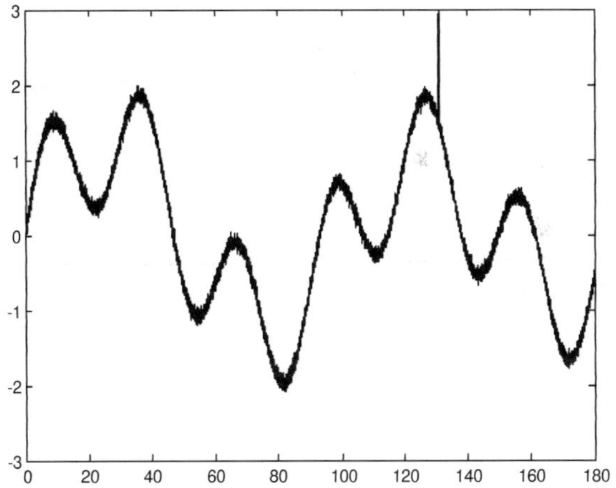

Figure 4. Outliers elimination result.

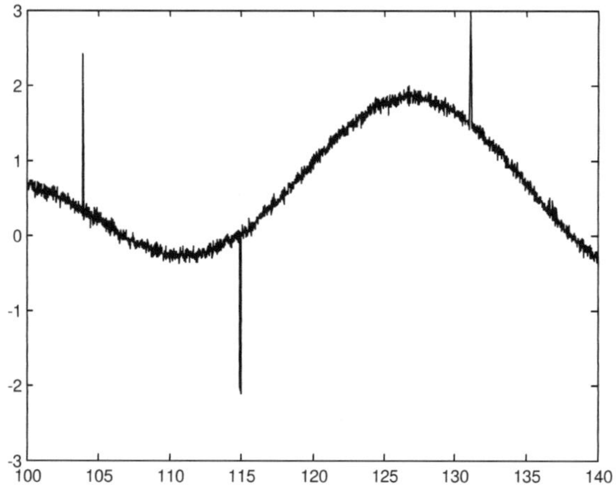

Figure 5. Initial data (interval between 100 s and 140 s).

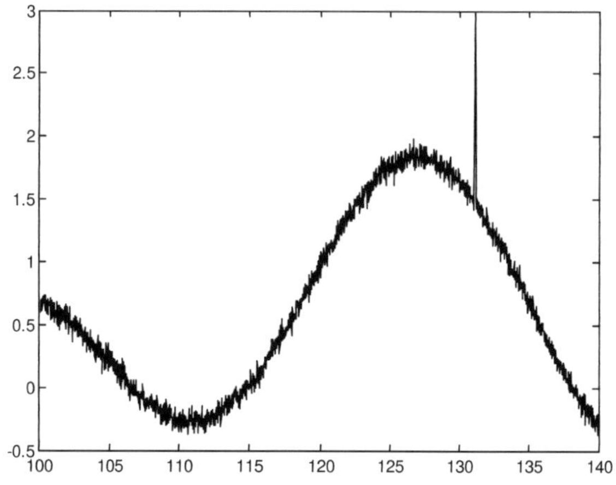

Figure 6. Outliers elimination result (interval between 100 s and 140 s).

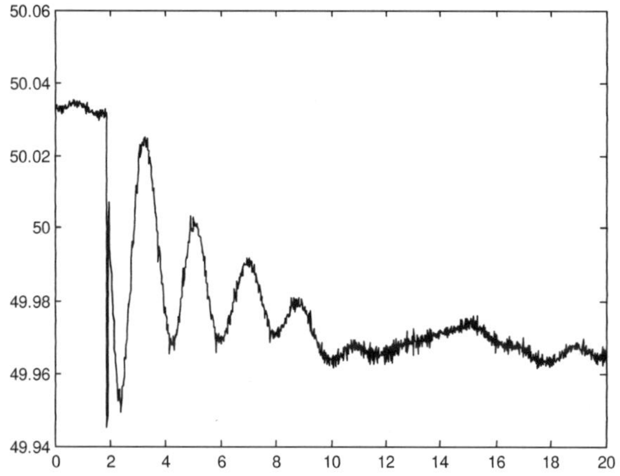

Figure 7. Actual WAMS data recorded after the disturbance.

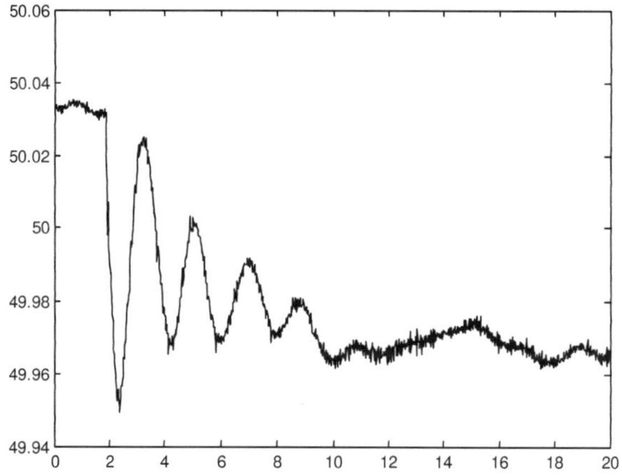

Figure 8. Outliers elimination result.

4. Conclusions

The algorithm is based on the concept of the irregular component of measurement signal for eliminating outliers has been developed for the purpose of data pre-processing before low-frequency oscillations automatic analysis and continuous monitoring. It has been proved to be effective and selective i.e. controlled by the adjustable parameter to the full extent. Since its application does not introduce significant calculations burden the algorithm may be implemented as a part of online monitoring tools.

5. References

[1] Zaidel A N 1965 *Measurement Errors Elementary Estimates* (Moscow: Science)

[2] Pazderin A V and Kochneva E S 2014 Bad data validation on the basis of a posteriori analysis *IEEE International Energy Conference ENERGYCON 2014 (Dubrovnik, Croatia)* pp 386–391

[3] Tavlintsev A S, Pazderin A V, Malozemova O Y and Chusovitin P V 2013 Identification of static polynomial load model based on remote metering systems information *13th*

International Conference on Environment and Electrical Engineering EEEIC 2013 (Wroclaw, Poland) pp 213–216

[4] Kochneva E and Sukalo A 2016 Impact of technical losses calculation method on bad data validation on the basis of a posteriori analysis *IEEE International Energy Conference ENERGYCON 2016 (Leuven, Belgium)* 7513899

[5] Kochneva E and Sukalo A 2016 Energy measurements verification based on evaluation residues *International Symposium on Industrial Electronics INDEL 2016 (Banja Luka, Bosnia and Herzegovina)* 7797779

[6] Chusovitin P V, Pazderin A V, Shabalin G S and Tashchilin V A 2014 Low-frequency oscillations identification in interconnected power system using PMU *Advanced Materials Research 860-863*, pp 2117–2121

[7] Berdin A S, Bliznyuk D I and Kovalenko P Y 2015 Estimating the instantaneous values of the state parameters during electromechanical transients *International Siberian Conference on Control and Communications SIBCON 2015 (Omsk, Russian Federation)* 7147001

[8] *State system for ensuring the uniformity of measurements. Metrology. Basic terms and definitions*, RMG 29-99

[9] Depablos J, Centeno V, Phadke A G and Ingram M 2004 Comparative testing of synchronized phasor measurement units *In proceeding of IEEE Power Engineering Society General Meeting*

[10] Mittelstadt W A, Hauer J F, Martin K E and Lee H 2005 *Evaluating the Dynamic Performance of Phasor Measurement Units: Experience in the Western Power System* Interim Report of August 2

[11] Kovalenko P Y, Berdin A S, Bliznyuk D I and Plesnyaev E A 2016 The flexible algorithm for identifying a disturbance and transient duration in power systems *International Symposium on Industrial Electronics INDEL 2016 (Banja Luka, Bosnia and Herzegovina)* 7797787

[12] Kovalenko P Y 2016 The extended frequency-directed EMD technique for analyzing the low-frequency oscillations in power systems *International Symposium on Industrial Electronics INDEL 2016 (Banja Luka, Bosnia and Herzegovina)* 7797788

Acknowledgments

The work was supported by Act 211 Government of the Russian Federation, contract № 02.A03.21.0006 and the Ministry of Education and Science of Russian Federation (in the framework of state assignment, № 13.1928.2014/K (project № 1928)).

IOP Conf. Series: Journal of Physics: Conf. Series **870** (2017) 012012 doi:10.1088/1742-6596/870/1/012012

Defining the parameters of a power transmission line equivalent circuit on the basis of phasor measurements

P Y Kovalenko[a] and A S Berdin

Automated Electrical Systems Department, Ural Federal University named after the first President of Russia B.N. Yeltsin. 19 Mira st., Yekaterinburg, Russia

[a]pkovalenko@urfu.ru

Abstract. Power system control is based on employing its computational model, the backbone of which is an equivalent circuit. The equivalent circuit is involved in solving the problems of steady-state computation and analysis, state estimation, transient analysis etc. Its elements parameters are generally defined by the corresponding equipment reference data or datasheets. Although considered to be constant, these parameters depend upon the element actual load, weather conditions and other factors. Consequently, the results of the outlined problems may be subject to significant bias due to the difference between the reference and the actual elements parameters. Hence the task of identifying the actual equivalent circuit parameters is of paramount importance. In terms of power system control the actual measurements are to be used in order to provide the relevant information on the considered power system component state. As for the transmission lines, the state measurements must include the currents and voltages at both ends of the line regarding the Π-shaped equivalent circuit. That said, at present time the equivalent circuit parameters might be defined involving modern systems of phasor measurements (WAMS) on a real-time basis. The method of defining the equivalent circuit parameters based on phasor measurements along with general relations between power system state parameters is proposed. It should be noted as well that while dealing with the actual WAMS data obtained from power system the measurement errors influence the results quite substantially.

1. Introduction

Most aspects of power system control rely on its computational model. The model, in turn, comprises the equivalent circuit for most power system components, parameters of which are traditionally defined according to reference values or datasheets and are generally assumed to be constant. Nevertheless, they depend upon actual operating and weather conditions, may be influenced by the adjacent equipment, may vary with equipment deterioration etc. As for the power transmission lines, the corresponding investigation [1] shows that the errors of defining their equivalent circuit parameters (longtitudal impedance $\underline{Z} = R + jX$ and shunt conductance $\underline{Y} = G + jB$) according to reference data is subject to significant errors, the results of the research are presented in table 1.

The transmission line equivalent circuit is involved in solving the problems of steady-state computation [2] and analysis, state estimation, transient analysis [3] and numerous other applications [4, 5]. One of the steady-state operation criteria is a static stability margin [6]. The stability limit of the line is generally derived from its operating conditions on the basis of its equivalent circuit. Excessive stability margins are undesirable due to several reasons: limited transmission capacity may result in extra capital

IOP Conf. Series: Journal of Physics: Conf. Series **870** (2017) 012012 doi:10.1088/1742-6596/870/1/012012

costs for additional generation or transmission capacity. Moreover, in case of power deficiency its price may significantly increase in certain area. At the same time insufficient stability margin limits the stability and reliability of the power system. Refining the equivalent circuit parameters allows to re-determine the stability margins [7] while ensuring the proper levels of stability and reliability.

Refining the equivalent circuit parameters has positive effect on electric power losses calculation since the process is based on the computational model as well.

Table 1. The errors of transmission line equivalent circuit parameters reference values.

Parameter	Reference value error range	Reference value error cause
R	–24…+8 %	Neglecting weather conditions and skin effect
X	–10…+8 %	Uncertainty of equivalent dimensions; multiply grounded lightning protection cables and parallel lines
G	25…30 %	Neglecting wire sag and grounded lightning protection cable effects, grounding impedance and air humidity
B	50…200 %	Construction design, operating and weather conditions

Precise parameters of transmission lines equivalent circuit have positive impact on transient analysis as the more accurate the model is – the more accurate results may be obtained by utilizing it.

In addition to the outlined fields the actual transmission lines parameters may find wide application for wires temperature control, fault location [8], defining automated protection systems thresholds [9] etc. Hence the problem of defining the valid equivalent circuit parameters is of crucial importance in terms of power system controllability, stability and reliability.

2. Defining the parameters of a transmission line equivalent circuit

Various researchers have investigated the problem at hand [7–11]. Solving the stated problem implies following one of the two approaches. The first of them is based on utilizing simultaneously the measurements of voltages and currents obtained from both ends of the line in two different system states [10]. Measurements in two states are involved since the line is considered as a quadrilateral regarding its Π-shaped equivalent circuit. The second approach presumes investigating the measurements of a single system state. Utilizing synchronized phasor measurements allows developing the algorithm for defining the transmission line parameters that relies on the fundamental relations between the voltages and currents at the ends of the transmission line. Moreover, various algorithms are being developed for the purpose of identifying the power system equipment parameters based on instantaneous electrical state measurements arrays [12–14] collected from, for example, digital fault recorders (DFRs) installed at power system entities.

3. The proposed method

Wide-area monitoring systems implementing the concept of synchronized phasor measurements enable obtaining the direct measurements of voltages and currents from both ends of the transmission line and also the derived values of active and reactive power. These parameters correspond to the set used in steady-state analysis software etc. so the actual measurements are convenient to compare with the computational models. Although it may be useful in particular applications, the three-phase equivalent circuit is rarely employed so the basic one-line equivalent circuit is considered from this point on, with few exceptions. The traditional Π-shaped equivalent circuit comprises a longtitudal impedance $\underline{Z} = R + jX$ and shunt conductance $\underline{Y} = G + jB$ and the measurements are distributed as shown in figure 1.

The following is correct for the equivalent circuit:

$$\dot{I}_{12} = \dot{I}_1 - \dot{U}_1 \cdot \underline{Y} = \dot{I}_2 + \dot{U}_2 \cdot \underline{Y}. \quad (1)$$

Consequently,

$$\underline{Y} = \frac{\dot{I}_1 - \dot{I}_2}{\dot{U}_1 + \dot{U}_2}, \quad (2)$$

and the equivalent circuit shown in figure 2 may be put in correspondence with equation (2).

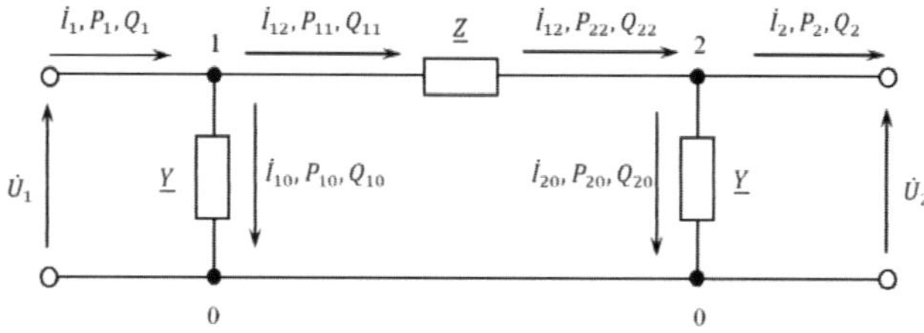

Figure 1. Π-shaped equivalent circuit and measurements.

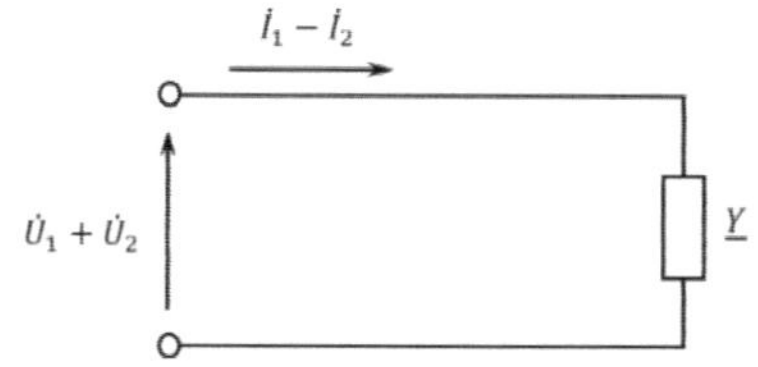

Figure 2. The equivalent circuit reduced to $\underline{Y}$ conductance.

That said, the shunt conductance $\underline{Y} = G + jB$ is obtained according to equation (2) and hence the longtitudal impedance $\underline{Z}$ is derived as

$$\underline{Z} = R + jX = \frac{\dot{U}_2 - \dot{U}_1}{\dot{I}_{12}} = \frac{\dot{U}_2 - \dot{U}_1}{\dot{I}_1 - \dot{U}_1 \cdot \underline{Y}}. \quad (3)$$

The proposed method has been validated involving a set of simulated data and employed in order to obtain the actual parameters of a transmission lines in operation, as described in the next section.

4. Method validation

The developed method validation includes two stages: validation involving simulated data and validation involving actual data collected from the WAMS system installed in the Unified Power System of Russia.

4.1. Validation based on a set of simulated data

The simulated data under consideration has been obtained by means of MATLAB software package (version R2009b) and Simulink module. The 500 kV transmission line connecting the buses Bus 1 and Bus 2 with $\underline{Z} = 5 + j59.66\Omega$ and $\underline{Y} = 2 - j880\mu S$ has been modelled. Two different operation states have been considered as well: the low-load and the high-load states with the corresponding power flows from Bus 1 to Bus 2 of $60 + j500MVA$ and $500 + j100MVA$. The simulation results are presented in table 2. As the line and the load are symmetrical all the measurements of voltages and currents are listed for phase A.

Similar to the simulation results, the equivalent circuit parameters are defined for a single phase. The derived parameters are $\underline{Z} = 5.021 + j59.679\Omega$, $\underline{Y} = 2.112 - j879.665\mu S$ and $\underline{Z} = 5.022 + j59.603\Omega$, $\underline{Y} = 2.182 - j877.926\mu S$ for two states, correspondingly. The errors appear to be less than 1 % for all of the parameters with the exception of ~10 % for active conductance.

It can be seen that the equivalent circuit parameters may be obtained directly from phasor measurements with practically appropriate accuracy which is a major step forward compared to the reference data.

IOP Conf. Series: Journal of Physics: Conf. Series **870** (2017) 012012 doi:10.1088/1742-6596/870/1/012012

Table 2. Simulation results.

	Low-load state ($60 + j500MVA$)	
	Voltage, V	Current, A
Bus 1	$293876.14\angle 0.06°$	$296.75\angle -77.79°$
Bus 2	$268538.90\angle -0.29°$	$540.93\angle -83.45°$
	High-load state ($500 + j100MVA$)	
	Voltage, V	Current, A
Bus 1	$293362.63\angle -0.01°$	$580.64\angle 7.13°$
Bus 2	$289111.05\angle -6.79°$	$589.67\angle -18.10°$

4.2. Validation based on actual WAMS data

The next stage of the developed algorithm validation involves actual data collected from the WAMS system deployed in the Unified Power System of Russia. The test cases of a single transmission line and two adjacent transmission lines were considered.

4.2.1. Single transmission line. First, the equivalent circuit parameters have been defined for the 500 kV transmission line of roughly 200 km length with 993 MW power flow. The line is equipped with the phasor measurement units (PMUs) at both ends. The considered measurements include voltages and currents phasors of three phases. The scheme of the PMUs installation is shown in figure 3. The results (for phases A, B and C) are shown in table 3. There are expected values (EV) and standard deviations (STD) for the time span of 10 minutes.

It can be seen that the situation is close to the one with the simulated data. The relative errors for the parameters are quite significant for active conductance and longtitudal resistance while reactive parameters are defined rather accurately. The significant errors are associated with the nature of the active conductance, as outlined in the Introduction, and the initial measurement errors. At this point it is worth noting that the measurement errors could not be corrected without measurements excessiveness and this leads to the next section where the two adjacent transmission lines are considered.

4.2.2. Two adjacent transmission lines. WAMS development allows to extend the algorithm of defining the transmission lines equivalent circuit parameters based on synchronized phasor measurements obtained not from the single line but from the group of lines. The measurements excessiveness is utilized in order to achieve higher precision of the defined parameters.

Two adjacent 750 kV transmission lines are considered in this section. The PMUs are installed at both ends of each line as shown in figure 4, the total of four 10-minute data sets was collected in order to run the algorithm. The results are shown in table 4. There are relative errors for the defined parameters as well.

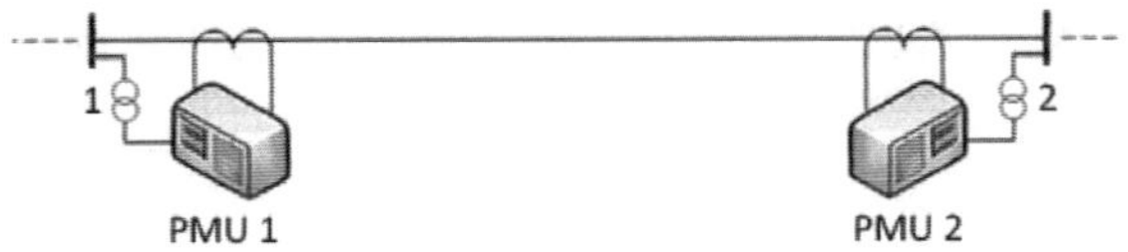

Figure 3. PMUs installation scheme for the single line case.

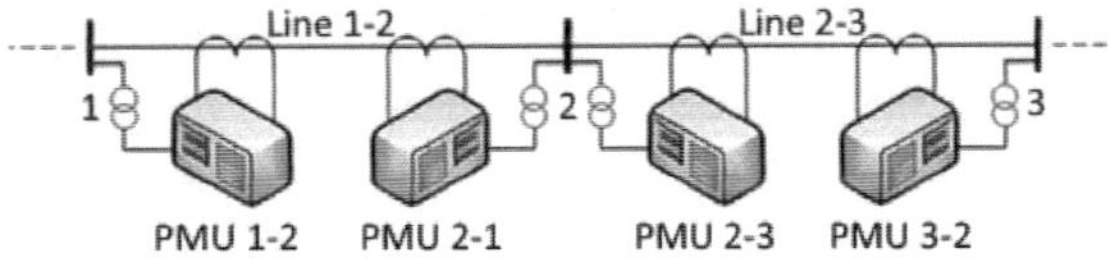

Figure 4. PMU installation scheme for the two adjacent lines case.

IOP Conf. Series: Journal of Physics: Conf. Series **870** (2017) 012012 doi:10.1088/1742-6596/870/1/012012

Table 3. The defined equivalent circuit parameters.

	EV and STD (for three phases)		
	A	B	C
R,Ω (5.42)	6.001 0.102	1.351 0.0855	1.272 0.101
X,Ω (68.30)	63.476 0.0519	65.050 0.0503	64.186 0.0508
G,S (7×10^{-6})	1.205×10^{-4} 1.870×10^{-6}	1.498×10^{-4} 1.429×10^{-6}	1.143×10^{-4} 1.366×10^{-6}
B,S (-9.830×10^{-4})	-9.445×10^{-4} 4.370×10^{-6}	-9.595×10^{-4} 4.374×10^{-6}	-9.132×10^{-4} 4.497×10^{-6}

Table 4. The defined equivalent circuit parameters.

Parameter (reference value)	The defined values for two lines (relative error, %)	
	Line 1-2	Line 2-3
R,Ω (9.8, 2.66)	487,555 (4875 %)	2,741 (3 %)
X,Ω (112, 35.34)	162,661 (45 %)	78,693 (123 %)
$G,\mu S$ (0, 0)	11552	49
$B,\mu S$ (–480, –477)	-1187 (-19 %)	-10956 (-2197 %)

Table 5. Initial measurements errors correction.

Parameter (reference value)	EV and STD (for three phases)			
	PMU 1-2	PMU 2-1	PMU 2-3	PMU 3-2
Voltage magnitude, %	0,50	-0,43	0,50	-0,46
Voltage angle, degrees	-0,48	0,45	0,45	0,50
Current magnitude, %	0,31	0,50	0,50	0,35
Current angle, degrees	0,50	-0,40	-0,39	0,50

Table 6. The defined equivalent circuit parameters.

Parameter (reference value)	The defined values for two lines (relative error, %)	
	Line 1-2	Line 2-3
R,Ω (9.8, 2.66)	35,211 (259 %)	3,118 (17 %)
X,Ω (112, 35.34)	181,320 (62 %)	40,857 (16 %)
$G,\mu S$ (0, 0)	1287	35
$B,\mu S$ (–480, –477)	-1885 (27 %)	-1012 (112 %)

According to the relative error values, only a few parameters may be considered correct. The relative errors significantly exceed the deviations of the actual parameters from their references listed in table 1. This allows for the conclusion that the result cannot be treated as the parameters refinement.

Although the results are quite poor, the measurements excessiveness enables automatic correction of initial data systematic errors. The criterion function of the parameters errors sum is minimized employing the nonlinear optimization approach. At that, the imposed limitations correspond to the instrument transformers rated accuracy of 0.5 class [15, 16]. The optimization results are shown in table 5. The equivalent circuit parameters defined after the initial data correction ("re-defined") are presented in table 6.

One can see that the relative errors have been significantly decreased and now they are close to the limits outlined in table 1. We also have to keep in mind that limiting the error by instrument transformer class 0.5 is quite optimistic since the actual errors may excess the rated values [17, 18]; in case the error is not limited in that way the accuracy of defining the equivalent circuit parameters reaches 10 %, which indicates the result may be treated as the reference parameters refinement.

5. Conclusion

The problem of defining the actual parameters of a transmission line equivalent circuit is relevant. The computational algorithm of appropriate accuracy has been developed and validated. The proposed algorithm may find extensive industrial application in various electrical engineering problems.

Evolution and expansion of modern measurement devices based on the concept of synchronized phasor measurements offer the potential to investigate the power system equipment and performance on a deeper level with the outcome of a more efficient and economical operation. The field of application of modern measurement systems includes but is not limited to defining the actual equipment parameters. Although the results of the proposed algorithm validation hold much promise, there are barriers, such as measurement errors and insufficient quantity of measurement devices, hindering the development in the outlined direction.

With this in mind the general suggestion for the PMUs to be installed is utilizing the instrument transformers of higher accuracy, which would allow to implement the algorithm of defining the transmission line equivalent circuit parameters. The secondary benefit would be the possibility to detect and correct the systematic measurement errors by solely software means.

6. References

[1] Idelchik V I, Novikov A S and Palamarchuk S I 1979 *Equivalent Circuit Parameters Errors in Power System Steady-State Calculations (Irkutsk)*

[2] Tavlintsev A S, Pazderin A V, Malozemova O Y and Chusovitin P V 2013 Identification of static polynomial load model based on remote metering systems information *13th International Conference on Environment and Electrical Engineering EEEIC 2013 (Wroclaw, Poland)* pp 213–216

[3] Chusovitin P V and Pazderin A V 2012 Implementation of power system model identification for locating in-phase generators *3rd IEEE PES Innovative Smart Grid Technologies Conference Europe (Berlin, Germany)* 6465721

[4] Kochneva E and Sukalo A 2016 Impact of technical losses calculation method on bad data validation on the basis of a posteriori analysis *IEEE International Energy Conference ENERGYCON 2016 (Leuven, Belgium)* 7513899

[5] Kochneva E and Sukalo A 2016 Energy measurements verification based on evaluation residues *International Symposium on Industrial Electronics INDEL 2016 (Banja Luka, Bosnia and Herzegovina)* 7797779

[6] *Methodological instructive regulations on power system stability* (Approved June 30th 2003 by the order 277 of RF Ministry of Energy)

[7] Klapper U, Kruger M and Kaiser S 2005 Reliability of transmission by means of line impedance and K-Factor measurement *In Proc. 18th International Conference and Exhibition on Electricity Distribution (Turin, Italy)*

[8] Shi D, Tylavsky D J, Logic N and Koellner K M 2008 Identification of short transmission-line parameters from synchrophasor measurements *In Proc. 40th North American Power Symposium (Calgary, Canada)*

[9] Kim Il-D and Aggarwal R K 2006 A study on the on-line measurement of transmission line impedances for improved relaying protection *Electrical Power and Energy Systems, vol 28, no 6* pp 359–366

[10] Khokhlov M V 2013 The errors of defining the transmission line equivalent circuit parameters on the basis of phasor measurements *The 6th International Scientific and Technical Conference «Power industry: Viewpoint of the Youth – 2013» proceedings (Novocherkassk, Russian Federation)*

[11] Liao Y and Kezunovic M 2009 Online optimal transmission line parameter estimation for relaying applications *IEEE Transactions on on power delivery, vol 24, n 1*

[12] Kolosok I N and Korkina E S 2014 Detecting systematic errors of instrument transformers according to state estimation approach on the basis of PMU measurements *International Exhibition and the XXII Conference "Relay Protection and Automation for Electric Power Systems" (Moscow, Russian Federation)*

[13] Ivano I E and Murzin A U Identifying the single-circuit power transmission line equivalent circuit parameters based on synchronized phasor measurements *5th International scientific and technical conference "Actual trends in development of Power System Relay Protection and Automation" (Sochi, Russian Federation)*

[14] Berdin A S, Bliznyuk D I and Kovalenko P Y 2015 Estimating the instantaneous values of the state parameters during electromechanical transients *International Siberian Conference on Control and Communications SIBCON 2015 (Omsk, Russian Federation)* 7147001

[15] *Current transformers. General specifications* International standard GOST 7746-2001 (Approved January 1st 2003)

[16] *Voltage transformers. General specifications* International standard GOST 1983-2001 (Approved January 1st 2003)

[17] Pazderin A V and Kochneva E S 2014 Bad data validation on the basis of a posteriori analysis *IEEE International Energy Conference ENERGYCON 2014 (Dubrovnik, Croatia)* pp 386–391

[18] Kovalenko P Y, Berdin A S, Bliznyuk D I and Plesnyaev E A 2016 The flexible algorithm for identifying a disturbance and transient duration in power systems *International Symposium on Industrial Electronics INDEL 2016 (Banja Luka, Bosnia and Herzegovina)* 7797787

Acknowledgments

The work was supported by Act 211 Government of the Russian Federation, contract № 02.A03.21.0006 and the Ministry of Education and Science of Russian Federation (in the framework of state assignment, № 13.1928.2014/K (project № 1928)).

Comparing the techniques of defining the synchronous machine load angle

P Y Kovalenko[1,a] and A N Moiseichenkov[2]

[1]Automated Electrical Systems Department, Ural Federal University named after the first President of Russia B.N. Yeltsin. 19 Mira st., Yekaterinburg, Russia
[2]Electrical Machines Department, Ural Federal University named after the first President of Russia B.N. Yeltsin. 19 Mira st., Yekaterinburg, Russia

[a]pkovalenko@urfu.ru

Abstract. The low-frequency oscillations are natural for power systems and may arise due to both small variations of load and large disturbance. The effect of slight load changes may significantly differ for cases of low-magnitude permanent oscillations, which may be considered acceptable, and unstable oscillations, which may lead to a major system emergency. The existing trend of increasing the capacity of long-range power transmission has led to the situation where inter-area oscillations may appear underdamped or even rising in terms of magnitude. Effective oscillations detection with the corresponding countermeasures along with eliminating the prerequisites leading to the oscillations is a guarantee of minimizing their negative consequences. Therefore, it is of crucial importance to perform continuous monitoring which is to provide the information on the "source" of oscillations – a generator or a group of generators, which do not contribute to the oscillations damping or even support their development. The algorithm of quantitative estimation of synchronous generators participation in low-frequency oscillations damping based on synchronized phasor measurements has been proposed previously. It implies utilizing the concept of synchronizing power as a measure of the capability of the machine to maintain synchronous operation. The load angle of the generator is necessary to define the value of the synchronizing power and since the direct measurement of the load angle is generally not available the techniques of its derivation have been developed. The comparison of these techniques is presented with the estimation of the adopted assumptions effect on the synchronizing power evaluation results.

1. Introduction

Low-frequency oscillations (LFO) are one of the important problems faced in power system around the world. The oscillations jeopardize power systems' reliability and stability, which leads to a significant decrease in transmission capacity. Therefore, the operation becomes less economic given the necessary levels of reliability and stability are ensured [1, 2].

The oscillations are naturally occurring in power systems and may result from minor load variations as well as disturbances such as generators or transmission lines trips. Large-scale centralized power systems interconnected by the weak tie-lines is a typical feature of the state-of-the-industry power engineering [3]. Another trend is distributed generation units integration with the resulting decrease of power system inertia constant and increasing sensitivity to small disturbances. In order to improve the controllability of power systems, modern equipment supplemented by fast-response control systems is introduced: asynchronized synchronous generators, FACTS (Flexible AC Transmission Systems), energy storage systems etc. Continuous changes in power system structure and parameters are other factors driving the LFO and threatening the equipment secure operation and the consumers power supply as well.

The resonance frequencies (eigen-frequencies) of the LFO in power systems fall within a range of 0.1÷3.0 Hz [4, 5]. The oscillations comprise local along with the inter-area ones, which may involve the whole power system. The local LFO characterize the mutual oscillatory movement of the synchronous machines within a power system region while the inter-area ones correspond to the mutual oscillations of the regions or machine groups against each other.

The analysis of power system state parameters often shows that there are time periods, within which the LFO damping is poor [6]. The presence of the underdamped oscillations means that the system operation may be unreliable due to the unpredictability of its dynamic response. Since all possible variations of operational conditions cannot be simulated using the dynamic model of a power system, the measurements-based monitoring of the LFO damping is crucial for preventing the potential threats before they may contribute to the development of a major emergency.

The LFO damping in power systems depends mostly upon control systems actions: automatic excitation and speed controllers, power system stabilizers [7], but may vary significantly subject to generators and load parameters and characteristics [8].

Ubiquitous introduction of Wide-Area Monitoring Systems (WAMS) based on the technology of synchronized phasor measurements provides the possibility to collect high-accuracy time-synchronized measurements of power system state parameters form Phasor Measurements Units (PMU) installed at power system entities. Moreover, advanced communication and data processing technologies allow to obtain these measurements "on-line" and to derive the LFO parameters, including frequency, magnitude and damping, on a real-time basis [9]. That said, the monitoring is to be performed not only during the transients or after the disturbances [10], but also during steady-state operation under normal conditions.

2. Estimation of synchronous generator participation in low-frequency oscillations damping

It is appropriate to adopt the quantity of power spent by the generator on the counteraction of the rotor slip relative to the system frequency as a measure of damping qualities of the generator [11]. This quantity is referred to as synchronizing power and the load angle is necessary to derive it. Three possible techniques of defining the load angle of a generator are described in detail in [12]. The selection of a technique follows a set of measured parameters with the measurements being based on on-site PMUs. The most accurate technique is obviously the one implying the load angle is directly measured mechanically by capturing the rotor angle and, optionally, angular velocity. In case the direct measurement is not possible, the load angle may be derived from the generator's electric parameters with a number of assumptions being adopted [12].

Under the circumstances of the derived load angle uncertainty, which is driven by the assumptions made, the problem of assessing the said uncertainty is relevant. The main aim of the uncertainty assessment is defining the possible field of application for the derived load angle value and determining the necessity to equip the generators with the load angle measurement systems.

Since the estimation of synchronous generator participation in low-frequency oscillations damping based on the phasor measurements and the synchronous machine computational model depends significantly on the assumptions adopted for the simulation, it is appropriate and rational to perform assessment for actual generators or their physical models.

3. Comparing the results of defining the synchronous machine load angle

In order to compare the techniques of defining the synchronizing power of a synchronous machine operating in generator mode, the experiments have been carried out on a physical model of a nonsalient-pole generator. The generator under consideration is a component of the "Digital-analog-physical Complex" operated by the Scientific and Technical Center of the Unified Power System (STC UPS).

The Digital-analog-physical complex includes the world's largest electrodynamic simulator (EDS) of electric power system (over 1000 models of physical generators, prime movers, transformers, transmission lines, complex load, direct current transmissions, FACTS, etc.) [13].

To implement the synchronizing power analysis algorithms, the EDS was equipped with the measurement system including the PMU capable of measuring the excitation parameters of a generator and the subsystem for measuring the rotor angle of a generator, comprising the magnet sensor and the tooth-wheel. The data obtained from the subsystem is recorded by the PMU as well. The measurement system was provided by Prosoft-Systems company [14], figure 1 shows the structure of the system. The fragment of the rotor angle measurement is presented in figure 2. The Prosoft-Systems SignW utility is employed for the raw data pre-processing. The electric parameters are derived from their instantaneous values of 10 kHz sampling rate in accordance with [15].

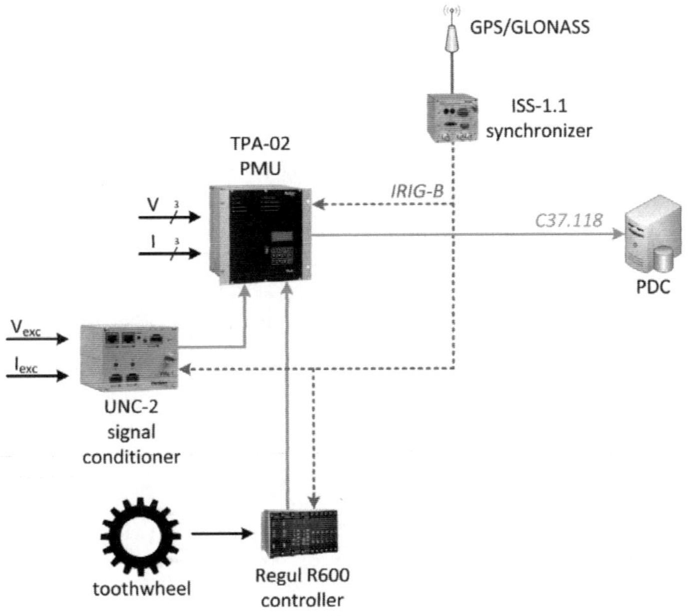

Figure 1. The structure of the measurement system.

The test data under consideration is the transient resulting from the load drop and surge represented by the corresponding changes of mechanical torque applied to the generator shaft. Figure 3 and figure 4 show the changes in generator's voltage frequency f and active power $P_{em} = P_g$ [11].

The load angle and the synchronizing power of the generator during the considered LFO were derived by three techniques [12]. The stator resistance was neglected due to its small value. Figure 5 shows the load angle calculation results. One can see that the derived values differ from the results of the direct measurements. It also should be noted that the result of direct measurements may be further processed to obtain the smooth $\theta(t)$ curve.

Prior to the derivation of the load angle by using the technique 3 [12] the inductance of the machine x is to be predefined with the value being refined according to the variations of frequency f. That said the predefined values may not correspond to the operational conditions preceding the disturbance. This disagreement may result in offsetting the derived values of the load angle. The following two values of the inductance were selected: the one derived experimentally on the basis of no-load and short-circuit characteristics of the machine ($x = x_d$) and the one derived on the basis of the directly measured load angle corresponding to the steady-state preceding the disturbance ($x = x_q$).

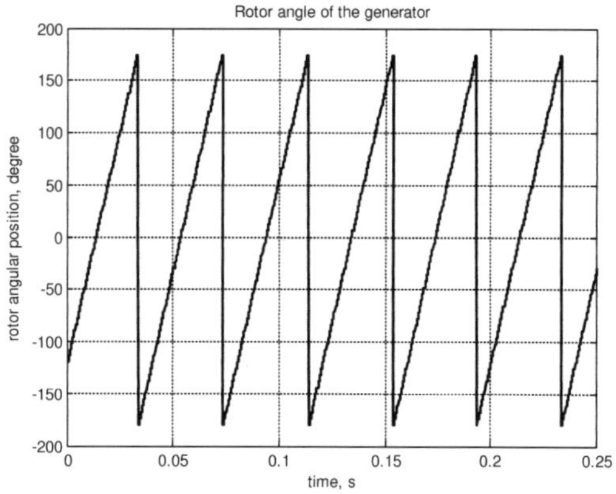

Figure 2. Rotor angle, degrees.

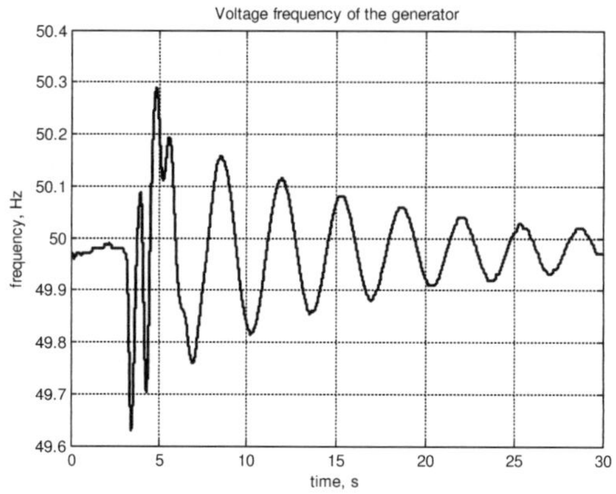

Figure 3. Voltage frequency f of the generator, Hz.

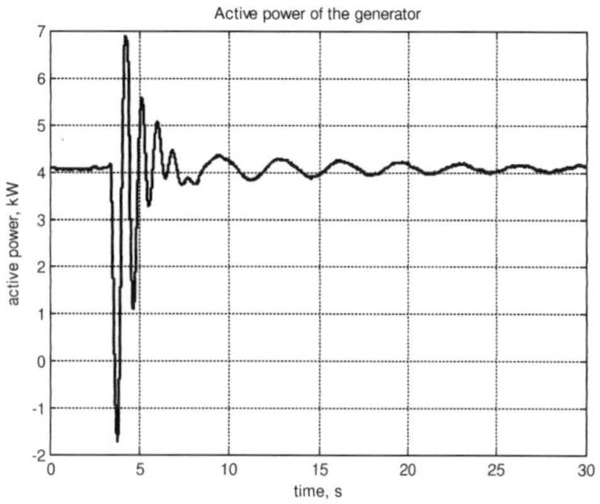

Figure 4. Active power of the generator, W.

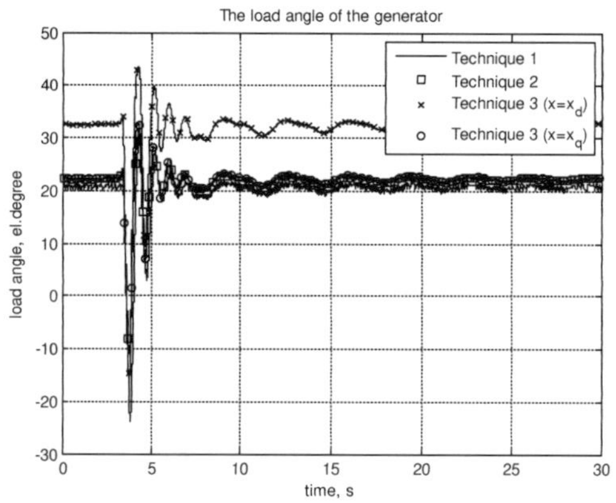

Figure 5. The load angle of the generator θ, el. degrees.

It is not necessary to predefine the inductance value during the derivation by using the technique 2 [12] since the stator electro-motive force (EMF) is defined on the basis of no-load characteristic of the machine. The no-load characteristic is recorded in advance with the rotor rotation speed $\omega_r(t)$ and excitation current $I_f(t)$ being measured. Consequently, this technique is recommended for the analysis of small disturbances.

Under the described circumstances the technique 1 is the most preferable to be used, which implies that the load angle is defined on the basis of the rotor angle measurements. If that method is adopted, the results of the load angle defining are less dependent upon the condition of the generator's magnetic core. Adopting this technique allows to derive P_s during either low-magnitude LFO or large disturbances.

The power-angle curve and the synchronizing power during the steady-state symmetrical operation may be defined according to the analytic formulas, developed under the assumption that the capacity of the system is infinitely larger than the capacity of the synchronous machine [11] given the state parameters and x_d, x_q are known. Accordingly, the $P(\theta)$ and $P_s(\theta)$ will vary with the changes of the machine operational conditions. During the transient the ($P(t), \theta(t)$) coordinates of the machine's operational point within $P-\theta$ axes are defined by the system conditions, the rotor's inertia constant, the machine's synchronizing and asynchronous damping torques and controllers' actions. Using the previously obtained results it is possible to plot the power-angle curve of the generator and its $P(t)$ against $\theta(t)$ trajectory. Figure 6 shows the $P(\theta)$ curve, corresponding to the steady-state prior to the disturbance, and the trajectory of the machine's operational point ($P(t),\theta(t)$). The load angle $\theta(t)$ is obtained using the technique 1.

The synchronizing power $P_s(\theta)$ may be derived as "quasistatic" – the value of the specific synchronizing power, i.e. tangent tilt of $P(\theta)$ at current machine's operational point, corresponding to its operational conditions at current time instant t. The value characterizes the capability of the machine to maintain synchronous operation given the state parameters do not vary; the asynchronous damping torque of the machine and the controllers' actions are neglected.

The synchronizing power $P_s(t)$ during the transient was derived for the possible values of the load angle θ obtained according to the three techniques. Figure 7 shows the results – the synchronizing power $P_s(t)$ derived by the analytic formulas developed under the adopted assumptions.

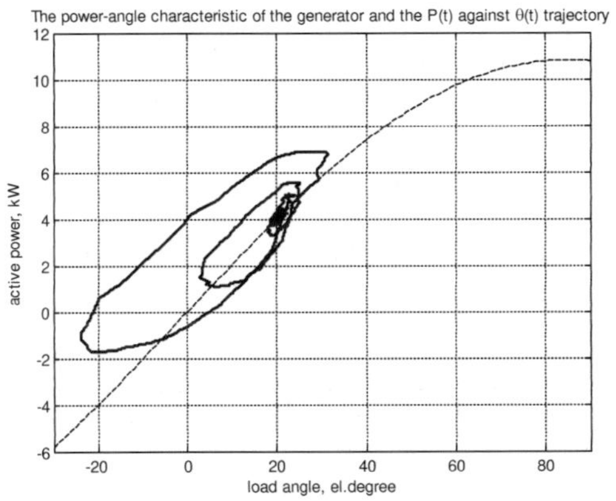

Figure 6. Power-angle curve $P(\theta)$ and the $P(t)$ against $\theta(t)$ trajectory.

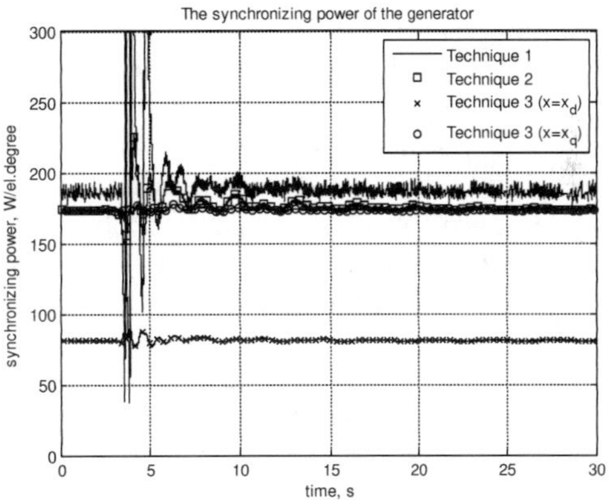

Figure 7. Synchronizing power of the generator $P_s(t)$, W/el. degree.

The alternative approach to calculation of the synchronizing power is numerical differentiation $\partial P(t)/\partial\theta(t)$. The direct differentiation allows to obtain the "dynamic" value of the synchronizing power $P_s(t)$ with consideration of the generator's rotor inertia constant, the changes in system operational conditions, the changes in the generator's voltage, voltage frequency and excitation current, the synchronizing and asynchronous damping torques of the machine and the actions of the controllers. Nonetheless, the result of the direct differentiation contains tangential discontinuities in the vicinity of the points where $\partial\theta \to 0$ and may turn out indefinite during quasi-steady-state operation, when $\begin{cases}\partial\theta \to 0 \\ \partial P \to 0\end{cases}$. Thus the further application of the obtained result is quite difficult. In order to obtain the practically applicable result the various numerical differentiation techniques may be employed.

4. Conclusion

The results of the carried out comparison for the considered physical model of a synchronous generator and the test disturbance may be summarized with the following conclusions.

- The considered techniques provide acceptable results.
- The most reliable one is the technique of the load angle defining on the basis of the rotor angle measurements. The computational methods may be employed in case the direct measurements

are missing, however their field of application is to be limited according to the errors resulting from the adopted assumptions.

- Estimation of a synchronous generator participation in the LFO damping on the basis of synchronizing power will improve the power system controllability and allow detecting the controllers' improper operation.

It is desirable to extend the comparative analysis of the load angle defining techniques involving the same physical model and different disturbances. It is of crucial importance to introduce as well the models comprising various generators and actual data obtained from power system entities.

The described methodology is expected to be implemented within the System of Power System Controllers Monitoring [16]. The main aim of the System is to detect in time the underdamped LFO and to define the source of the oscillations. At the time the algorithms of the System allow to identify the conventional faults of the excitation system of the generators: lack of or untimely excitation forcing, misoperation of the minimum/maximum excitation current limiter.

5. References

[1] Ayuev B I 2008 *Methods and Models for Efficient Operational Control of United Power System of Russia* D.Sc. thesis in engineering science (Yekaterinburg, Russian Federation)

[2] Chusovitin P V, Pazderin A V, Shabalin G S and Tashchilin V A 2014 Low-frequency oscillations identification in interconnected power system using PMU *Advanced Materials Research 860-863*, pp 2117–2121

[3] Pavlushko S A 2012 Synchronous Generators Automatic Excitation Regulation as an Effective Instrument for Providing Reliable Synchronous Operation of Generating Equipment and the Unified Energy System of Russia in general *Power Plants no 7* (Moscow, Russian Federation)

[4] Messina A R 2009 *Inter-area Oscillations in Power Systems* (Springer US)

[5] Rogers G 2000 *Power Systems Oscillations* (Springer US)

[6] Chusovitin P V and Pazderin A V 2012 Implementation of power system model identification for locating in-phase generators *IEEE PES Innovative Smart Grid Technologies Conference Europe (Berlin, Germany)* 6465721

[7] Chusovitin P V, Pazderin A V, Shabalin G S and Tashchilin V A 2015 PSS tuning method based on power system model identification using PMU *IEEE PES Innovative Smart Grid Technologies Conference Europe (Istanbul, Turkey)* 7028817

[8] Tavlintsev A S, Pazderin A V, Malozemova O Y and Chusovitin P V 2013 Identification of static polynomial load model based on remote metering systems information *13th International Conference on Environment and Electrical Engineering EEEIC 2013 (Wroclaw, Poland)* pp 213–216

[9] Kovalenko P Y 2016 The extended frequency-directed EMD technique for analyzing the low-frequency oscillations in power systems *International Symposium on Industrial Electronics INDEL 2016 (Banja Luka, Bosnia and Herzegovina)* 7797788

[10] Kovalenko P Y, Berdin A S, Bliznyuk D I and Plesnyaev E A 2016 The flexible algorithm for identifying a disturbance and transient duration in power systems *International Symposium on Industrial Electronics INDEL 2016 (Banja Luka, Bosnia and Herzegovina)* 7797787

[11] Ivanov-Smolenskiy A V 2004 *Electrical Machinery: High School Student Textbook for Those Studying at Certified Specialists Training Course* 2nd ed (Moscow: MEI)

[12] Berdin A S, Zakharov Y P and Kovalenko P Y 2014 Estimation of synchronous generator participation in low-frequency oscillations damping based on synchronized phasor measurements *WIT Transactions on Ecology and the Environment vol 190: Energy Production and Management in the 21st Century (2 Volume Set) vol 1* pp 319–325

[13] Scientific and Technical Center of Unified Power System, *Digital-analog-physical Complex the Unique Test Area of System Operator of the United Power System,* https://www.ntcees.ru/departments/nio_3/dep_nio_3_en.php, https://www.ntcees.ru/departments/nio_3/en/Eelektrodinamics_buklet.pdf

[14] Prosoft-Systems, *TPA-02-PMU – Microprocessor based device of synchronized phasor measurements*, http://www.prosoftsystems.ru/products_eng/product-tpa.htm

[15] Berdin A S, Bliznyuk D I and Kovalenko P Y 2015 Estimating the instantaneous values of the state parameters during electromechanical transients *International Siberian Conference on Control and Communications SIBCON 2015 (Omsk, Russian Federation)* 7147001

[16] Gerassimov A S, Esipovich A Kh, Sheskin E B, Shtefka Y, Shukov A V and Negreev A P 2013 The results of integrated and field testing of the System of Power System Controllers Monitoring, *Proceedings of the 4th international scientific and technical conference "Actual trends in development of power system relay protection and automation" (Yekaterinburg, Russian Federation)*

Acknowledgments

The work was supported by Act 211 Government of the Russian Federation, contract № 02.A03.21.0006 and the Ministry of Education and Science of Russian Federation (in the framework of state assignment, № 13.1928.2014/K (project № 1928)).

IOP Conf. Series: Journal of Physics: Conf. Series **870** (2017) 012014 doi:10.1088/1742-6596/870/1/012014

Influence of measurement levels number on the accuracy of calculated estimate during the electrical energy measurements verification

E S Kochneva[1,a] and A Sukalo [2]

[1]Automated Electrical Systems Department, Ural Federal University named after the first President of Russia B.N. Yeltsin, 19, Mira str, Yekaterinburg, Russia
[2]Elektroperenos – Elektroprijenos BiH, Banja Luka, Bosna and Hercegovina

[a]e.kochneva@prosoftsystems.ru

Abstract. Verification of electric energy measurement is a very important issue. The research deals with the modified method of testing equations, which enables measurement-based electrical energy estimation of higher accuracy compared to the initial measurements. The influence of the testing equations number on the estimation error value is investigated. Test results are presented for 14-node IEEE scheme. The necessity of using the limited measurement level number is proved.

1. Introduction

Information about the measured values of electrical energy received from Automatic Meter Reading (AMR) systems is generally quite reliable. Such systems use modern microprocessor-based electrical energy meters as well as reliable modern methods of measurements data retrieve, transmission and storage. Nevertheless, the possible failure of any system component will cause data errors or gaps, which are difficult to detect and eliminate. Mathematical approaches to measurements errors identification are the most promising and require the least financial investment [1, 2].

Current trend in electrical energy automated metering systems is their further technological advancement. Although the systems and their component devices become more and more technically complex, the actual electrical energy losses value remains high and significantly exceeds the technical and regulatory-specified values. Grid companies draw up the electrical energy balance monthly, with the balance serving as the basis for deriving the actual energy losses in order to compare them to the regulatory-specified ones. In case the commercial losses value is significant, the grid company suffers financial damage.

The causes of excessive commercial losses include underestimation of energy supplied to consumers, and intentional theft. The supplied energy underestimation is often driven by the instrument transformers overload. The AMR complex measurement error depends upon the feeder operation conditions and load with energy value typically being underestimated under off-nominal conditions. Hence, negative AMR complexes systematic errors might lead to commercial energy losses increase. Another issue is intentional commercial metering data distortion, especially in case the metering complexes are owned by consumers, who are interested in measurements understatement to decrease the electricity costs.

IOP Conf. Series: Journal of Physics: Conf. Series **870** (2017) 012014 doi:10.1088/1742-6596/870/1/012014

Commercial losses have been increasing during over the last 20 years, resulting in financial damage to energy exchange parties and grid companies, in particular. Commercial component of the overall energy losses doubles the expenses of transmission companies as they bear the financial responsibility. First, grid company receives less than due for energy transport because of commercial losses. And after that it pays for commercial losses during the actual energy losses covering procedure.

Commercial losses at 0.4 kV level are the most expensive for grid companies as energy transport rates corresponding to that level are the highest. Grid companies energy audit experience shows that the highest share of commercial losses corresponds to the low-voltage levels as well, which results in energy transport tariff increase, leading, in turn, to electricity cost rise for all end-users, especially for low-voltage ones.

Decreasing the commercial share of energy losses is possible through locating their sources, identifying the causes and eliminating them. Energy measurements verification enables locating the major commercial losses sources in the network [3]. In order to effectively establish the financial relations between market participants exact and reliable information about energy exchange is necessary. Since measurements in AMR systems are obtained at predefined regular intervals of time, it is possible to employ mathematical models and methodology to assess the system performance as well as verify and validate the commercial measurements on a real-time basis [4].

2. Energy measurements validation

State estimation theory, being widely used in power engineering to verify telemetry data [1, 2, 5], is characterized by high degree of scientific development [6–12]. Body of mathematics employed in state estimation solution includes fundamental steady-state equations along with computational optimization techniques.

Initial data in state estimation problem includes network diagram and its parameters; impedances and conductances in the diagram are assumed to be constant. The solution involves estimating the voltage magnitude and angle of each node and currents of all brunches as well as nodes' active and reactive power injections.

A wide variety of crucial problems might be solved based on the obtained data: identifying the invalid measurement data and locating its source; estimating the parameters, that are not measured; assessing the measurement errors values and their statistical properties; obtaining the estimates of higher accuracy compared to the actual measurements. The approaches and methods, allowing to solve the listed problems, are well-developed and have found wide application in automated dispatch management systems algorithms. Hence, they might as well be adapted and applied to the challenge of bad data validation in electric energy measurements.

During previous research [13–15] the impossibility of employing the traditional steady state equations for energy flow calculation was shown and confirmed. Implementation of these equations leads to significant simulation errors. The problem of obtaining the energy flows and energy losses corresponding to all network elements based on energy measurements was named the Energy Flow problem. It was proposed to adopt the node and branch energy balance equations as energy flow state equations system:

$$\begin{aligned} &\sum_{j\in\omega_i} W_{ij} + W_i = 0, \quad i = 1,2\ldots N; \\ &W_{ij} + W_{ji} - \Delta W_{ij} = 0, \quad ij = 1, 2,\ldots M, \end{aligned} \tag{1}$$

where W_i is the i node energy injection; W_{ij} is the i node connections energy flow; ω_i is the array of the nodes adjacent to the i node; N is the number of nodes in the network; M is the number of branches in the network. This approach has a lot in common with the traditional methodology of actual and admissible imbalances comparison.

IOP Conf. Series: Journal of Physics: Conf. Series **870** (2017) 012014 doi:10.1088/1742-6596/870/1/012014

3. Measurements estimated equivalents calculation, maximum admissible errors

Within the framework of state estimation theory the equations containing all measured variables are referred to as testing equations. The testing equations approach can be applied to energy measurements in energy flow problem [4, 17]. Testing equations system might be formed by excluding all parameters, which are not measured, from energy flow state equations system by any available mathematical means. Substitution of all measurements in case they could contain no errors into the testing equations system would lead to zero residual error. Therefore, the measurements errors drive non-zero residual error, while bad data results in significant residual error.

The number of testing equations determines the number of a single energy flow measurement levels. Thus, redundant measurements presence enables obtaining a range of estimated equivalents for the same variable. The estimate accuracy (maximum admissible error) can be derived according to

$$\delta_{W_i} = \sqrt{\sum_{j=1}^{n} a_j^2 \cdot \delta_{W_j}^2}\,, \tag{2}$$

where a_j is the coefficient of W_i in the i testing equations number. The less measurements are included into the corresponding TE the lower the estimate maximum admissible errors.

Figure 1 presents the possible levels of W_1 energy flow metering. The first stage is forming the energy flow state equation system. The number of equations equals three: two equations for node energy balances and one for branch balance. The second stage is obtaining the testing equations system, which implies elimination of all variables, that are not measured, from the state equations system. The network shown in figure 1 is characterized by full measurement coverage; hence, the testing equations system is entirely in accord with the energy flow state equation system. The third stage is constructing the verifying equations group out of the testing equations system. The estimated energy flow is to remain on the left side of the equation and the rest variables are transferred to the right part. Three testing equations allow to obtain three verifying equations for the estimated energy flow:

$$\begin{aligned} W_1 &= W_2, \\ W_1 &= W_3 - \Delta W, \\ W_1 &= W_4 + W_5 - \Delta W. \end{aligned} \tag{3}$$

Energy flow value here can be calculated based on the three levels of measurement, i.e. this energy flow is measured at three levels. Local redundancy coefficient [13] is calculated as the number of redundant measurements plus one. At least one measurement is required to provide observability for figure 1 diagram [16]. The number of measurements in considered example is four (measurements 4 and 5 form one node measurement).

Assuming the relative accuracy of all AMR complexes to be 1.6 % [18], the relative accuracy of measurement levels 1 and 2 is also 1.6 % according to (2). On the supposition that the technical losses calculation error is 10 % and the value of technical losses are about 1 % of the energy flow under normal operation conditions, the relative accuracy of technical losses calculation is 0.01 %. Therefore, the relative accuracy of the third and the forth levels measurements are, correspondingly, $\sqrt{1.6^2 + 0.01^2} \approx 1.603\%$ and $\sqrt{1.6^2 + 1.6^2 + 0.1^2} \approx 2.265\%$. The most accurate measurement estimated equivalent can be obtained by minimizing the relative measurement errors squares weighted sum:

$$F = \sum_{k=1}^{K_{red}} \frac{1}{(\delta_k \cdot W_k^{meas})^2} (W_k^{meas} - W^{est})^2 \to \min\,. \tag{4}$$

The summand number under the sum sign (4) equals K_{red}. δ_k values are calculated according to (2). The first factor $(\delta_k \cdot W_k)^{-2}$ in (4) corresponds to the relative accuracy of energy flow measurement. In case all the redundant measurements are the same, the function (4) can be reduced to a common denominator:

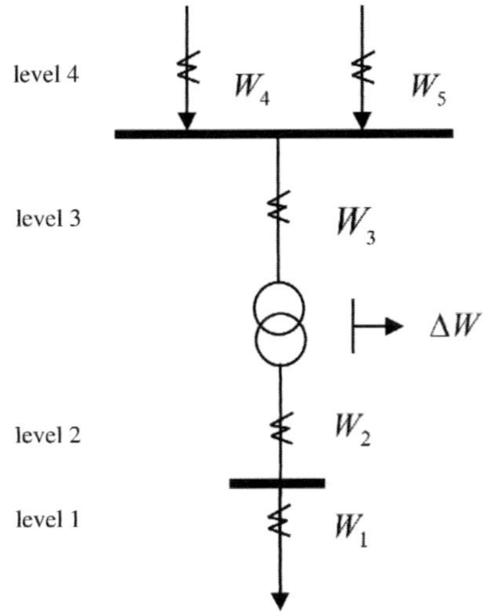

Figure 1. Measurements level

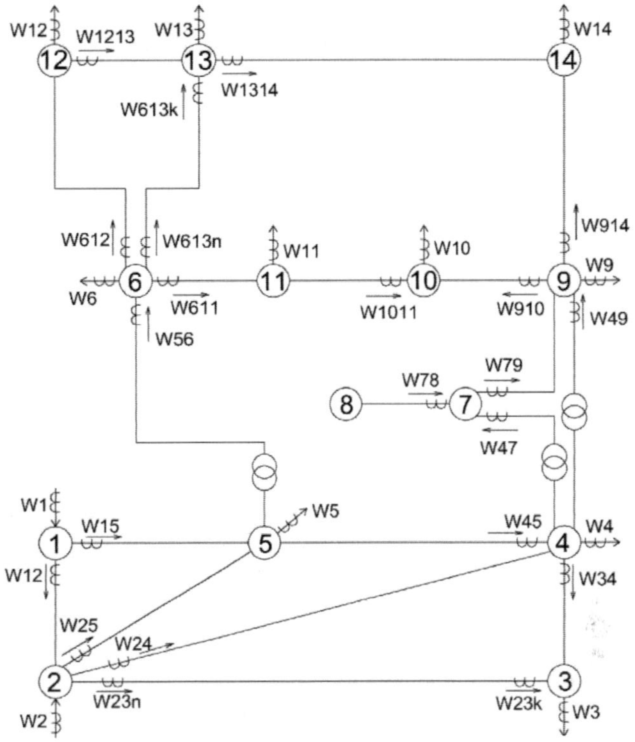

Figure 2. Considered test network.

$$F = \frac{\sum_{k=1}^{K_{red}}\left(\prod_{k=2}^{K_{red}}\delta_k^2\right)}{\prod_{k=1}^{K_{tot}}\delta_k^2}\cdot\left(\frac{W^{est}}{W^{meas}}\right)^2 + C\ . \qquad (5)$$

The relative error of the energy flow measurement estimated equivalent based on the redundant measurements can be evaluated in accordance with

$$\delta_{est} = \left[\prod_{k=1}^{K_{red}}\delta_k^2\cdot\left(\sum_{k=1}^{K_{red}}\prod_{k=2}^{K_{red}}\delta_k^2\right)^{-1}\right]^{0,5}. \qquad (6)$$

In the considered example the energy flow estimate relative error based on the remaining redundant measurements would be as low as $\delta_{W_i}\approx 0.856\%$. The resulting relative error of energy flow estimate is almost half as large as the AMR complex W_1 measurement relative error itself, assumed to be 1.6 %. If the technical losses calculation error is not taken into account, the estimate relative accuracy will be $\delta_{W_i}\approx 0.855\%$. In practice the technical losses calculation error can be as high as 20 % and even more. However, this level of the losses value uncertainty affects the estimation relative error insignificantly. In the considered network under the condition of 20 % losses calculation error, the estimation relative error is still 0.858 % and not more than 0.869 % even if the technical losses calculation error increases up to 50 %.

4. Test example, estimation of the measurement levels optimal number

IEEE 14-node scheme presented in figure 2 is considered as a test example, it includes 110 kV and 220 kV voltage levels.

IOP Conf. Series: Journal of Physics: Conf. Series **870** (2017) 012014 doi:10.1088/1742-6596/870/1/012014

Table 1. Calculated estimates and verifying equations errors.

Parameter	Energy flow value, MW·h	Admissible error, MW·h	Admissible error, %	Estimate error, %
$W_1^{level(1)}$	17.879	0.26104	1.460	1.460
$W_1^{level(2)}$	18.237	0.190	1.044	1.044
$W_1^{level(3)}$	17.947	0.188	1.046	0.739
$W_1^{level(4)}$	16.705	1.069	6.399	0.734
$W_1^{level(5)}$	16.478	1.066	6.472	0.730
$W_1^{level(6)}$	16.310	1.067	6.540	0.725
$W_1^{level(7)}$	16.796	1.070	6.369	0.720
$W_1^{level(8)}$	17.611	2.668	15.147	0.720
$W_1^{level(9)}$	15.297	2.664	17.412	0.719
$W_1^{level(10)}$	15.632	2.765	17.685	0.718
$W_1^{level(11)}$	16.315	2.762	16.927	0.718
$W_1^{level(12)}$	14.567	3.640	24.990	0.717
$W_1^{level(13)}$	15.431	3.784	24.521	0.717

The installed AMR complexes are marked in simplified form by the windings of instrument transformers with reasonable measurement redundancy being provided. A variety of operation conditions was simulated employing RASTR software complex in order to obtain the resulting energy distribution. The relative measurement error limit of all AMR complexes installed is assumed to be 1.46 %, all measurements are assumed to correspond to that predefined error limit. Herein W_1 measurement is investigated; the verifying equations are formed for it. 13 measurement levels are considered.

Table 1 shows the W_1 energy flow estimates, MW·h, and admissible error values, MW·h and on a percentage base. The right column of table 1 gives the estimate admissible errors obtained on the basis of corresponding measurement levels number. It is convenient to illustrate the results as a plot of estimate admissible error against measurement levels number, as shown in figure 3. It is clear from the graph that the admissible error after level 4 slightly changes, hence it is sufficient to use level 5 for the case under consideration.

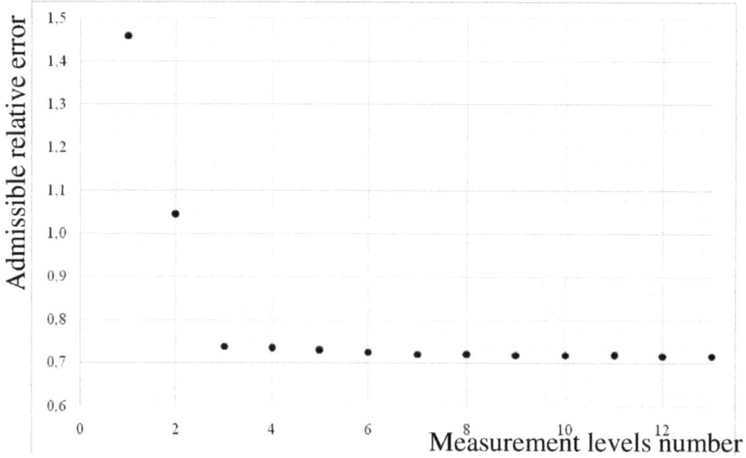

Figure 3. Admissible relative error as a function of measurement levels number.

5. Conclusion

Modern AMR systems are quite reliable sources of data comprising electrical energy flows corresponding to network elements. However, any system component fault or malfunction might result in significant measurement errors or bad data occurrence. Mathematical approach to identifying those issues is the least demanding in financial terms. State estimation theory methodology adapted to fit the energy flow problem requirements enables successfully addressing the issue of electrical energy measurements verification. Testing equations method modification allows obtaining energy measurements estimated equivalents, admissible errors of which are lower than of the initial measurements. Herein, the technical energy losses evaluation error influences the estimation accuracy insignificantly. Provided four measurement levels, the estimate error is decreased twice compared to the error of the measurement itself. Thus, the estimate accuracy depends upon measurement redundancy: the number of redundant measurements determines the testing equations number and the estimate admissible error value. However, involved measurement levels number increase influences the computational burden and equations system complexity. The optimum measurement levels number for the test case under considered conditions equals five.

6. References

[1] Monticelli A, Wu F F, Yen M, Multiple Bad Data Identification for State Estimation by Combinatorial Optimization. *Proceesings of the PICA Conference*, pp. 452-460, (1985)

[2] Nian-De X, Shi-Ying W, Er-Keng Y, A new approach for Detection and Identification of Multiple Bad Data in Power System State Estimation. *IEEE Transactions on Power Systems*, **PAS-101**, no. 2, pp.454-462, (1982).

[3] Pazderin A, Samoylenko V, Localization of non-technical energy losses based on the energy flow problem solution. *Proceedings of the 6th IASTED Asian Conference on Power and Energy Systems, AsiaPES 2013*, pp.100-103, (2013).

[4] Egorov A, Kochneva E, Pazderin A, Detection of systematical errors of AMR system complexes, *Advanced Materials Research* **960-961**, pp 1342-46, (2014).

[5] Gamm A, Kolosok I, *Bad data detection in measurements in electric power system*. Nauka, Novosibirsk, Sib. Enterpr. RAS, (in Russian), (2000).

[6] Schweppe F, Douglas J, Rom D, Power System State Estimation, Part I - Exact Model, *IEEE Transactions on Power Apparatus and Systems*, **PAS-89**, pp. 120-125, (1970).

[7] Schweppe F, Douglas J, Rom D, Power System State Estimation, Part II - Approximate Model, *IEEE Transactions on Power Apparatus and Systems*, **PAS-89**, pp. 125-130, (1970).

[8] Monticelli A, Electric Power Syaytem State Estimation. *Proceedings of the IEEE*, vol. 88, no.2, pp. 262-282, (February 2000).

[9] Korres G, A distributed multi-area state estimation, *IEEE Transactions on Power Systems*, **26**, no. 1, pp. 73-84, (2011)

[10] Wu F F. Power System State Estimation: A Survey, *International Journal of Electric Power and Energy Systems*, **12, no. 2**, pp. 80-87, (1990)

[11] Chusovitin P, Polyakov I, Pazderin A, Three-phase state estimation model for distribution grids. *Proceedings of International Conference on the Science of Electrical Engineering ICSEE*, (2016).

[12] Abur A, Exposito A G, *Power System State Estimation*. Marcel Dekker inc., New York, (2004)

[13] Plesniaev E, Pazderin A, Data acquisition system faults detection. *Proceedings of IEEE Conference on Contril Applications*, pp.1390-94, (2003).

[14] Pazderin A, Kokin S, Egorov A, Kochneva E, Solution of energy flow problem using state estimation technique, *ECON Proceedings (Industrial Electronics Conference)*, pp. 1736-41.

[15] Plesniaev E, Pazderin A, Analysis of the computation techniques for energy flow problem solving. *Proceedings of the International Conference on Computer as a Tool II EUROPCON*, pp.1469-1472, (2005).

[16] Pazderin A, Egorov A, Eroshenko S. The energy meters allocation in electric systems on the basis of observability theory, *Proceedings of 9th Conference on Environment and Electrical Engineering, EEEIC*, pp. 167-170, (2010).
[17] Pazderin A, Kochneva E, Bad data validation on the basis of a posteriori analysis. *Proceedings of the IEEE International Energy Conference ENERGYCON*, pp. 386-391, (2014).
[18] RD 34.09.101-94. *Typical instruction for electricity metering in its generation, transmission and distribution*. M. ORGRES, S. (in Russian) (1995).

Acknowledgments
The work was supported by Act 211 Government of the Russian Federation, contract № 02.A03.21.0006 and the Ministry of Education and Science of Russian Federation (in the framework of state assignment, №13.1928.2014/K (project №1928)).

IOP Conf. Series: Journal of Physics: Conf. Series **870** (2017) 012015 doi:10.1088/1742-6596/870/1/012015

Testing equations method for electrical energy measurements systematic errors detection and measurements results adjustment

E S Kochneva[1,a] , A V Pazderin[1] and A Sukalo[2]

[1]Automated Electrical Systems Department, Ural Federal University named after the first President of Russia B.N. Yeltsin, 19, Mira str, Yekaterinburg, Russia
[2]Elektroperenos – Elektroprijenos BiH, Banja Luka, Bosna and Hercegovina

[a]e.kochneva@prosoftsystems.ru

Abstract. State estimation theory is widely used for telemetry verification. Similar approaches can successfully be applied to electric energy measurements. Testing equations method was proposed in the frameworks of the state estimation theory and adopted for bad energy measurements detection. It is proposed to employ the modified testing equations method for detecting the systematic measurements errors of automatic electrical energy meter reading complexes. It is shown and confirmed that provided the statistical sample it is possible to detect and evaluate the systematic error, which, in turn, allows taking this error into account and adjusting the measurements results.

1. Introduction. Problem relevance

Economic efficiency is currently given the most attention in almost every industry field, and power engineering is not an exception. Scope of responsibility of grid companies includes electrical energy transportation from generation facilities to consumers. Hence, their main source of income is the fee received for energy transport. Electrical energy measurements obtained by means of energy metering systems constitute the basis of financial relations between the parties within wholesale and retail electricity markets.

Actual measurements sources are energy metering complexes installed at the borders of responsibility areas of various energy exchange participants. Due to electrical energy tariffs (i.e. prices for consumers) differentiation on an hourly basis, the automatic meter reading (AMR) systems are being widely deployed, expanded and advanced at a fast pace. AMR systems are intended to timely provide reliable electrical energy measurements data in order to properly establish the financial settlements between the energy exchange parties.

However, electrical energy measurements, besides typical errors, may contain bad data resulting from significant metering devices measurement errors, exceeding the admissible rates, as well as measurements data transformation distortions or communication channels noise and disturbances [1]. Energy flows measurements validity and accuracy monitoring is of paramount importance for any AMR system operation, which allows minimizing the financial risks for energy exchange participants in case of bad data occurrence within commercial measurements.

IOP Conf. Series: Journal of Physics: Conf. Series **870** (2017) 012015 doi:10.1088/1742-6596/870/1/012015

As a matter of fact, electrical energy measurements do not reflect the true energy flow value corresponding to the metering complex installation point. Energy measurement, similar to any other one, can be generally represented by the sum of three quantities: true energy flow value W_i^{true}, which is actually unknown, and two measurement error components – systematic ζ_{Wi} and random σ_{Wi} errors:

$$W_i^{meas} = W_i^{true} + \zeta_{Wi} + \sigma_{Wi}. \quad (1)$$

Since the true energy flow value is unknown, its statistical parameters are, consequently, unknown, as well. Hence, it is of particular interest to estimate all values appearing in the expression (1) in order to improve measurements data reliability.

It is common knowledge that the closer the measured value is to the true one – the higher is the measurement accuracy. For this requirement to be met it is necessary that the metering complex measurement error would not exceed the sum of systematic and random errors, which constitutes the criterion of electrical energy measurement validity.

As the statistical properties of measurement errors are unknown, the concept of maximum admissible relative measurement error of electrical energy metering complex was introduced in metrology practice. This figure corresponds to the limits of the measured parameter valid values range. The probability of the measurement value to fall within this range is 95 %.

2. Systematic errors detection and measurements results adjustment

In power engineering the state estimation theory is widely applied for the purposes of detecting bad data, evaluating errors as well as solving various other problems regarding measurements data verification [2–8].

Metrological techniques of detecting the excessive measurement errors of electrical energy metering complexes are quite difficult to implement on a practical level, especially when dealing with high-voltage installations. Therefore, applying the mathematical approach in order to detect the metering complexes, the measurement error of which exceeds the regulatory-specified limits, is prospective and promising [9].

Modern electrical energy metering systems provide sufficient measurement redundancy. State estimation theory application allows to choose the optimal allocation of AMR-complexes to provide observability and sufficient redundancy level [10]. Figure 1 show the most common metering system structure from 110 kV busbar downwards.

Level I of the test subnetwork metering system shown in figure 1 is the 110 kV busbar input – 110 kV lead-in. Level II comprises the 6/10 kV busbars input – 6/10 kV lead-in. Level III features 6/10 kV busbars feeders output. Level IV is formed by the total lower-level substations input, corresponding to each 6/10 kV busbar feeder. Level V incorporates the equivalent lower-level substations feeders output.

Measurements redundancy offers the possibility to derive a single energy flow based on different measurements sets, which enables obtaining several calculated estimated equivalents corresponding to each actual measurement expressed as a linear aggregate of other measurements [11].

One of numerous possible approaches to the construction of estimated equivalents expressions is the topological method.

The first stage is selecting the measurement, for which the estimated equivalents will be obtained; W_{13} is taken by way of example for the following examination. Then topology-based balance edges are formed in such a way that the desired energy flow could be derived from them. Figure 2 illustrates a series of balance edges plotted by dash lines. Technical losses of electrical energy values corresponding to the network elements crossed by an edge are added to the calculated estimate expression. Herein, the losses calculation method is not strictly determined and the selection of one is defined by a set of initial data available. The most commonly used method is the technical losses calculation based on the considered period average power.

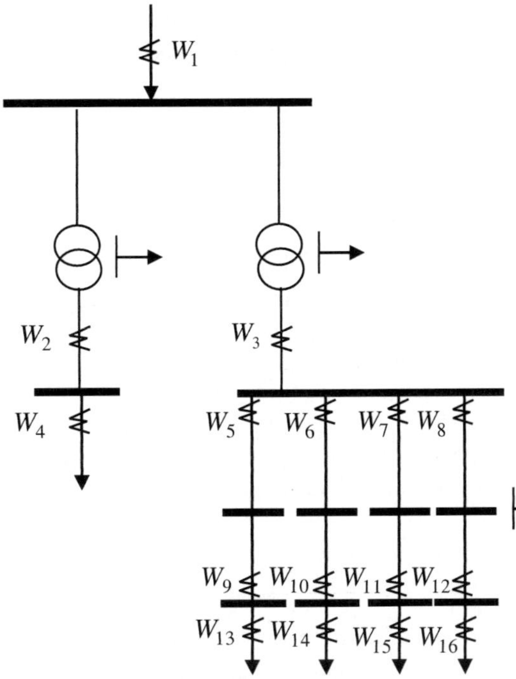

Figure 1. AMR system structure diagram

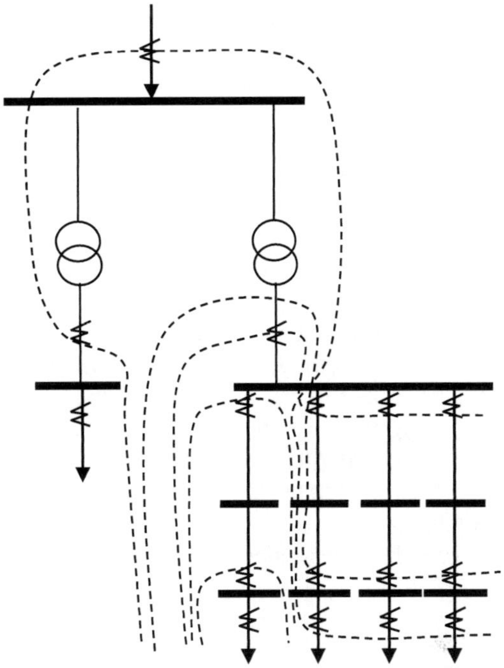

Figure 2. Balance edges corresponding to measurement 13.

The algebraic approach to the construction of similar expressions, worth noting as a more formalized one, was described in detail in [11–12].

State estimation theory is widely used to solve wide variety of electric power industry problems [13-14].

Several estimated equivalents are attributable to a single measured energy flow value W_{13} with the exact number being established on the basis of available redundancy level [11]. At that, the accuracy of the measurement itself accords with its metering complex accuracy rating.

After that the measurement calculated estimate is obtained from numerous estimated equivalents. First, the relative accuracy of W_i energy flow estimated equivalents, derived from testing equations, is to be defined.

The relative accuracy of each verifying equation is calculated according to

$$\delta_{W_i} = \sqrt{\sum_{j=1}^{n} a_j^2 \cdot \delta_{W_j}^2}\ , \tag{2}$$

where δ_{W_i} is the i metering compex accuracy rating.

The calculated estimate of energy flow value is the following:

$$W^{est\_total} = \frac{\sum_{i=1}^{N} \frac{1}{D_i} W_i^{est}}{\sum_{i=1}^{N} \frac{1}{D_i}}. \tag{3}$$

Hence, the measured energy flows values W_i^{meas} are used as the initial data for the purpose of calculating their estimated equivalents W_i^{est}.

Deriving the measurement estimated equivalent based on the verifying equations system also allows to define the measurements errors corresponding to each time period under consideration.

The higher is the metering complexes accuracy – the less the difference ΔW between the measured W_i^{meas} and the estimated W_i^{est} energy flow values is to be. Therefore, the effective metering complex measurement error might be evaluated based on the relative estimation residual value $r^{est} = \left(\Delta W_i / W_i^{meas} \right) \cdot 100\%$. The the criterion of measurement validity is the expression

$$r_i^{est} \le \delta_{\lim_i}, \quad (4)$$

where $\delta_{\lim_i}$ is the maximum admissible measurement error of metering complex i.

Previous work [11] propose to apply the modified testing equations method for demanding energy measurements verification. Modern AMR-systems have quite voluminous archives of the measured electric energy on the specific time intervals to provide a sufficient statistical sample.

It is possible to accumulate the statistical parameters of every metering complex estimated errors as far as multiple time periods are available for analysis. The subsequent statistics processing is performed in order to detect and evaluate the systematic error of every single metering complex. The metering compelex systematic measurement error is determined by the estimation residuals series expected value.

Hence, the metering complex proper functioning criterion is the following inequation:

$$\bar{r}_i^{est} \le \delta_{\lim_i}, \quad (5)$$

where $\bar{r}_i^{est}$ is the relative estimation residuals series expected value corresponding to the W_i measurement; $\delta_{\lim_i}$ is the maximum admissible measurement error of metering complex i.

3. Case study. Method validation based on the test network

The equivalent circuit was formed for the test subnetwork and two sets of operation conditions different in terms of consumers load values were defined and computed by means of the specialized software package.

All metering complexes accuracy ratings were calculated under the assumption of their compliance with RD 34.09.101-94 requirements [13]:

$$\delta_W = \pm 1.1\sqrt{\delta_I^2 + \delta_U^2 + \delta_m^2 + \delta_l^2}, \quad (6)$$

where δ_I and δ_U are the measurement errors of, correspondingly, voltage and current instrument transformers (GOST 7746-2001 and GOST 1983-2001), %; δ_m is the meter basic relative measurement error, %; δ_l is the error driven by the voltage loss on a connection line between the meter and the voltage instrument transformer, %.

The first three components in the metering complex maximum admissible error expression (6) were assumed to equal 0.5, and the voltage loss value of 0.25 % was adopted, which correspond to the accuracy ratings of the most common commercial metering complexes components in operation. The resulting maximum admissible relative error follows from here to be [16]

$$\delta_W = \pm 1.1\sqrt{0.5^2 + 0.5^2 + 0.5^2 + 0.25^2} = 1.134 .$$

The method was validated on the basis of the 24-hour time period, of which the two sets operation conditions were simulated during 12 hours each.

The estimation expressions system involves the electrical energy losses corresponding to the test subnetwork equivalent circuit components. It is proposed to calculate the losses based on the mean power value during the analyzed time period. Table 1 shows the comparison of the accurate losses, defined as the difference between the simulated energy flow values at the beginning and at the end of the transmission lines, and the losses calculated based on the mean power value, with the power flows being impacted by simulated measurement errors within the rated accuracy range and mean voltage values and equivalent circuit parameters being used.

Table 1. Energy losses values comparison.

$N_{beginning}$	N_{end}	Accurate losses	Losses based on mean power value	Relative error, %
1	2	0.01	0.01	-3.84
1	3	0.00	0.00	13.72
3	4	0.04	0.03	-16.26
3	5	0.19	0.15	-17.65
3	6	0.03	0.02	-31.12
3	7	0.10	0.10	-0.47
4	8	0.00	0.00	0.00
5	9	0.00	0.00	0.00
6	10	0.00	0.00	0.00
7	11	0.00	0.00	0.00

700 30-minute simulations were performed for the purpose of testing the systematic error detection method.

In practice it is rather difficult to accurately calculate the technical losses value for a time period due to the following reason. Technical losses comprise two components: load-based and semi-constant. The primary uncertainty source is the calculation of corona losses, which are heavily dependent upon environment conditions – temperature and humidity.

Moreover, in case the load-related losses calculation is based on the element resistance, its value is not exactly known as well. Under typical operation conditions the losses level commonly amounts for

Table 2. Verifying equations calculation results and their errors.

Verifying equation number	Estimate	Error, kW·h	Error, %
1	3.384	0.038	1.134
2	3.323	0.038	1.134
3	3.398	0.045	1.322
4	3.443	0.452	13.134
5	3.362	0.479	14.244
6	3.435	0.457	13.298
7	3.507	0.465	13.264
8	3.354	0.484	14.414
9	3.499	0.470	13.424
10	3.426	0.492	14.355
11	3.418	0.496	14.522
12	2.432	1.213	49.868
13	2.351	1.239	52.715
14	2.424	1.217	50.219
15	2.496	1.226	49.110
16	2.343	1.239	52.896
17	2.488	1.230	49.449
18	2.415	1.252	51.856
19	2.407	1.257	52.216

1–5 % relative to the element flow, hence the maximum losses calculation error is considerably lower than the metering complex maximum admissible error.

The losses calculation error of 20 % is adopted for the test case. The calculation is done based on the accurate losses value at the first stage.

19 verifying equations were constructed for W_{13} measurement using the algorithm described in [10] with the error in kW·h and on a percentage base being defined. The resulting values are listed in table 2.

The calculated estimate of the energy flow value is obtained using (3): $W_{13} = 3.363$. The energy flow estimated eqivalent maximum admissible error in the considered exmple is $\delta_{est} = 0.67\%$, which is almost half as large as the metering complex measurement error. The actual relative difference between the estimate and the true values equals 0.10 % while the difference between the measured and the true values reaches 0.70 %.

The similar approach applied to the losses based on the mean power value revealed the insignificance of the losses calculation method for the final outcome.

The same calculations were performed for 700 half-hour periods. The systematic of error of 13 % was intentionally included into the measurements series. Part of the results are shown for illustrative purposes in table 3.

The average estimation residuals obtained for all considered periods equals 11.5 %, which is close to the predefined systematic error.

The estimation residuals series expected value is the evaluation of the metering complex measurement error constant component. Therefore, the described methodology provides the possibility to adjust every metering complex measurements without expensive and time-consumig metrological calibration procedures. Furthermore, this method enables detecting the metering complexes requiring the exceptional calibration, which makes it unnecessary to calibrtate all metering complexes. However, it is impossible to adjust the commercial electrical energy maesurements by means of introducing the correction indexes due to the lack of legal framework, i.e. the regulations in this field.

The accumulated statistics and estimated equivalents calculations based on the verifying equations method allows to evaluate the systematic error of any single metering complex. This results in potential decrease of regular calibration costs by means of detecting the metering complexes, characterized by measurement errors exceeding the admissible value.

The statistical sample sufficiency is determined by the measurement interval length. In modern AMR-systems carrying out such calculations in online mode together with the previous results will allow monitoring the current state of the concrete AMR-complex.

The method allow to optimize the list of current metrological verification.

4. Conclusion

Methods of measurements verification based on the testing equations system showed good performance considering the test subnetwork. The testing equations method enables detecting the measurements containing bad data.

Verifying equations might be constructed on the basis of the testing equations system for the critically important commercial electrical energy measurements. Utilizing the verifying equations method allows to verify the key measurements in order to reduce the grid companies financial risks.

Table 3. Verifying equations calculation results and their errors.

Parameter	Period 1	Period 2	Period 3	Period 4	.....	Period 700
W_{13}^{meas}, kW·h	0.113339	0.113565	0.113452	0.111870		0.044974
δ^{meas}, %	1.13	1.13	1.13	1.13		1.13
$W_{13}^{est\_total}$, kW·h	0.099353	0.100785	0.098752	0.099769		0.039740
r^{est}, %	12.3	11.3	13.0	10.8		11.2

Systematic error evaluation gives the opportunity of adjusting the metering system measurement data.

References

[1] Pazderin A, Samoylenko V, Localization of non-technical energy losses based on the energy flow problem solution. *Proceedings of the 6th IASTED Asian Conference on Power and Energy Systems, AsiaPES 2013*, pp.100-103, (2013).

[2] Abur A, Exposito A G, *Power System State Estimation.* Marcel Dekker inc., New York, (2004)

[3] A. Monticelli, *State Estimation in Electric Power Systems – a Generalized Approach.* Norwell, M A: Kluwer, (1990)

[4] Schweppe F, Douglas J, Rom D, Power System State Estimation, Part I - Exact Model, *IEEE Transactions on Power Apparatus and Systems*, **PAS-89**, pp. 120-125, (1970)

[5] Schweppe F, Douglas J, Rom D, Power System State Estimation, Part II - Approximate Model, *IEEE Transactions on Power Apparatus and Systems*, **PAS-89**, pp. 125-130, (1970)

[6] Monticelli A, Electric Power Syaytem State Estimation. *Proceedings of the IEEE*, vol. 88, no.2, pp. 262-282, (February 2000)

[7] Monticelli A, Wu F F, Yen M, Multiple Bad Data Identification for State Estimation by Combinatorial Optimization. *Proceedings of the PICA Conference*, pp. 452-460, (1985)

[8] Nian-De X, Shi-Ying W, Er-Keng Y, A new approach for Detection and Identification of Multiple Bad Data in Power System State Estimation. *IEEE Transactions on Power Systems*, **PAS-101, no. 2**, pp.454-462, (1982)

[9] Gamm A, Kolosok I, *Bad data detection in measurements in electric power system.* Nauka, Novosibirsk, Sib. Enterpr. RAS, (in Russian), (2000).

[10] Pazderin A, Egorov A, Eroshenko S. The energy meters allocation in electric systems on the basis of observability theory, *9th Conference on Environment and Electrical Engineering, EEEIC 2010* 5489950, pp. 167-170, (2010).

[11] Kochneva E, Pazderin A, Sukalo A, Testing equation method modification for demanding energy measurements verification. *Proceedings 1st International Conference on Measurement Instrumentation and Electronic, ICMIE*, (2016).

[12] Pazderin A, Kochneva E, Bad data validation on the basis of a posteriori analysis. *Proceedings of the IEEE International Energy Conference ENERGYCON 2014*, pp. 386-391, (2014).

[13] Berdin A, Bliznyuk D, Kovalenko P, Estimating the instantaneous values of the state parameters during electromechanical transients, *International Siberian Conference on Control and Communications, SIBCON 2015 – Proceedings* 7147001 (2015).

[14] Kovalenko P, Berdin A, Bliznyuk D, Plesnyaev E, The flexible algorithm for identifying a disturbance and transient duration in power systems, *Proceedings of International Symposium on Industrial Electronics, INDEL*, (2016).

[15] Plesniaev E, Pazderin A, Analysis of the computation techniques for energy flow problem solving. *Proceedings of the International Conference on Computer as a Tool II EUROPCON 2005*, pp.1469-1472, (2005).

[16] RD 34.09.101-94. *Typical instruction for electricity metering in its generation, transmission and distribution.* M. ORGRES, S. (in Russian) (1995)

Acknowledgments

The work was supported by Act 211 Government of the Russian Federation, contract № 02.A03.21.0006 and the Ministry of Education and Science of Russian Federation (in the framework of state assignment, №13.1928.2014/K (project №1928)).

IOP Conf. Series: Journal of Physics: Conf. Series **870** (2017) 012016 doi:10.1088/1742-6596/870/1/012016

Implementation in an FPGA circuit of Edge detection algorithm based on the Discrete Wavelet Transforms

Issam Bouganssa[1], Mohamed Sbihi[1] and Mounia Zaim[1]

[1] Laboratory of System Analysis, Information Processing and Integrate Management, High School of Technology SALE, Mohammed V University in Rabat MOROCCO.

issam.bouganssa@gmail.com

Abstract. The 2D Discrete Wavelet Transform (DWT) is a computationally intensive task that is usually implemented on specific architectures in many imaging systems in real time. In this paper, a high throughput edge or contour detection algorithm is proposed based on the discrete wavelet transform. A technique for applying the filters on the three directions (Horizontal, Vertical and Diagonal) of the image is used to present the maximum of the existing contours. The proposed architectures were designed in VHDL and mapped to a Xilinx Sparten6 FPGA. The results of the synthesis show that the proposed architecture has a low area cost and can operate up to 100 MHz, which can perform 2D wavelet analysis for a sequence of images while maintaining the flexibility of the system to support an adaptive algorithm.

1. Introduction

Detecting edges within an image is an important characteristic of the image-searching process depending on the content. Faced with a large number of edge detection techniques and filters, it may be difficult to choose the most suitable approach for a specific collection of images, especially that there is no optimal technique for all the cases [1]. Moreover, it is crucial to study the characteristics of the images as a whole (synthetic or real image, fuzzy or clear, noisy) and according to their components (zones of high frequency versus low frequency) and then determine the most optimal edge detection approach in the selected set of detectors.

Edge extraction is used in the majority of cases, segmentation, image search depending on the content, identifying specific objects in images, or reconstructing objects in three dimensions, for this kind of recognition or reconstruction application [2]. Extracting the Edges is therefore an indispensable pre-treatment of the image before implementing other treatments.

In this paper, we use the Haar discrete wavelet transform for edge extraction to isolate the edge of the rest of the image in the horizontal, vertical and diagonal direction [3]. The algorithm is implemented in a Xilinx Spartan6 FPGA device that provides good performance of the integrated circuit platform for research and development. Indeed, FPGA technology has become an alternative for the implementation of software algorithms.

The results implemented in FPGA are displayed continuously on a screen via a VGA controller. The image will be stored in a FPGA block memory. The chosen technology allows real-time processing of image sequences while maintaining the flexibility of the system to support an adaptive algorithm.

2. Principe of contours

2.1. Frequency definition of contours

If the image is considered as a 2D signal, it is possible to pass in the frequency domain (by Fourier transform or wavelet for example) [4]. In this case, a contour can be seen as representing the high frequencies of the signal.

2.2. Spatial or temporal definition of contours

A contour can be seen as an abrupt change in the intensity of the image. There are several types of variations. A very similar way to the one quoted above is to consider the contours as a difference on the color. The contour detection filters this based on:

- Derivative first of a contour: gradient, Derivative operators of the first order, search for a passage through a maximum, or extra local maxima.
- Second derivative of an outline: Laplacian, Derivative operators of the second order, search for a passage by zero.

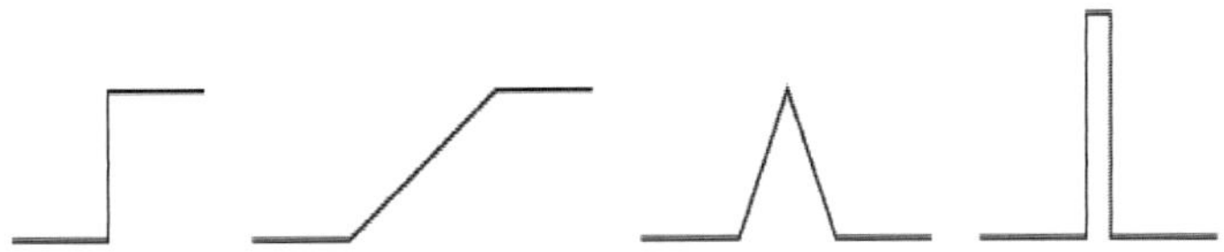

Figure 1. Contour profiles, Walking or ramp, roof, peak

2.3. Characteristics of contours

As well as the visual characteristics of color and texture, contours also have their peculiarities, to predict which detector will be the most effective. In order to estimate the efficiency of the detectors, we will refer to some errors encountered when detecting contours (Perfect contour Figure 2a):

- Omission of certain pixels on the contour to be detected. It is measured by counting the number of forgotten pixels with respect to the total number of pixels of the ideal contour (Figure 2b).
- Multiple responses by detecting multiple contours. It is measured by counting the number of ambiguous pixels compared to those that are not ambiguous (Figure 2d).
- Location: this error occurs when a pixel of an unambiguous ideal outline is not in the right place. It is measured by counting the total distance between the detected contour and the ideal contour. (Figure 2c).
- Sensitivity: This error is often related to noise and corresponds to false contours detected near the ideal contour. It is measured by counting the number of false contours and the total number of contours detected (figure 2e).

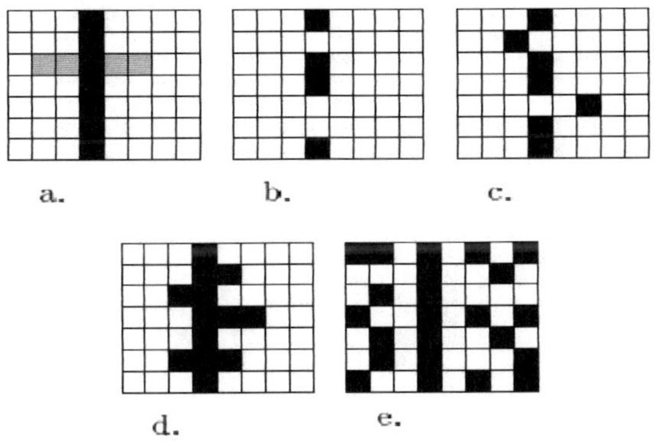

Figure 2. Examples of contour errors:
a: ideal contour,
b: Omission,
c: relocation,
d: multiple responses,
e: sensitivity.

IOP Conf. Series: Journal of Physics: Conf. Series **870** (2017) 012016 doi:10.1088/1742-6596/870/1/012016

3. Wavelets

Most real-world signals are not stationary, and it is precisely in the evolution of their characteristics (statistics, frequency, temporal, spatial) that most of the information they contain is located. Voice signals and images are exemplary.

3.1. Fourier Transform

The Fourier analysis (equation 1) proposes a global signal approach, integrations are made from minus infinity to infinity [3,4], and any notion of temporal (or spatial) localization for images disappears in the space of Fourier.

$$T^{fourier} f(\omega) = \int_{+\infty}^{-\infty} f(t)e^{-j\omega t}\, dt \quad (1)$$

It is therefore necessary to find a compromise, a transformation which informs the frequency content while preserving the location in order to obtain a representation of the time / frequency signal.

3.2. Fourier transform with sliding window

The first solution that comes naturally to mind is to limit the domain of temporal (or spatial) integration by means of a "window" function that can be dragged to explore the signal[5]; we obtain the Fourier transform with sliding window (equation 2).

$$T^{f\,glisse} f(t,\omega) = \int_{+\infty}^{-\infty} f(s)g(s-t)e^{-j\omega t}\, ds \quad (2)$$

In this case, it is understood that the analysis is not ideal, because if a low temporal resolution is automatically linked to the detection of the low frequencies, the detection of the high frequency components of the signal can be done with a higher temporal resolution. The two resolutions must vary in the opposite direction, keeping a constant product for an energetically regular paving of the time-frequency space.

3.3. Wavelet transform

The Wavelet Transform Allows a rational use of the time-frequency space by the realization in all cases of the best possible compromise between the temporal and the frequency resolution [6].

This principle is specified in equation 3:

$$T^{ond} f(a,b) = \frac{1}{\sqrt{a}}\int_{-\infty}^{+\infty} f(t)\psi\left(\frac{t-b}{a}\right) dt \quad (3)$$

In this expression, a is the scaling factor and b is the translation parameter. The variable a plays the role of the inverse of the frequency: More a is small less the wavelet is extended temporally, so the center frequency of its spectrum will bi higher.

This expression can also be interpreted as a projection of the signal on a family of analytic functions $\psi_{a,b}$ constructed from a "mother" function ψ according to equation 4:

$$\psi_{a,b}(t) = \frac{1}{\sqrt{a}}\psi\left(\frac{t-b}{a}\right) \quad (4)$$

It will be noted that the norm is preserved when the scaling factor is changed (equation 5):

$$\|\psi_{a,b}\|^2 = \int_{-\infty}^{+\infty} \frac{1}{a} \left|\psi\left(\frac{t-b}{a}\right)\right|^2 dt = \mathrm{cc}\int_{-\infty}^{+\infty} |\psi(x)|^2\, a dx = \|\psi\|^2 \quad (5)$$

We can therefore: $T^{ond} f(a,b) = \langle f, \psi_{a,b} \rangle$

The spatial-temporal resolution is calculated in the same way as before: If the temporal "width" of ψ (the standard deviation) is taken as a unit: σ = 1 then the "width" of $\psi_{a,0}$ can be calculated with equation 6:

$$\sigma_t^2 = \int t^2 |\psi_{a,0}(t)|^2 dt = \int t^2 \frac{1}{a} \left|\psi\left(\frac{t}{a}\right)\right|^2 dt = \int a^2x^2 \frac{1}{a} \langle\psi(x)\rangle^2 a dx \quad (6)$$

Which gives $\sigma_t = a$ it is also possible to calculate the frequency occupancy of the wavelet by calculating the standard deviation for the Fourier transform $\psi_{a,01}$ of $\psi_{a,0}$ taking as unit the standard deviation of the Fourier transform of the mother wavelet ψ equation 7:

$$\sigma_\omega^2 = \int \omega^2 |\widehat{\psi_{a,0}}(\omega)|^2 d\omega = \int \omega^2 \frac{1}{a} |a\widehat{\psi}(a\omega)|^2 d\omega = \int \frac{\xi^2}{a^2}\frac{1}{a} |a\widehat{\psi}(\xi)|^2 \frac{d\xi}{a} \quad (7)$$

IOP Conf. Series: Journal of Physics: Conf. Series **870** (2017) 012016 doi:10.1088/1742-6596/870/1/012016

We find $\sigma = \frac{1}{a}$ So that the elementary block in the time-frequency space has a constant surface while the temporal resolution is proportional to a and the frequency resolution is inversely proportional to a as seen in figure 3.

4. Principle of edge detection by wavelets

4.1. The types of wavelets use in 2D signal processing

The first wavelets used (apart from the **Haar** wavelet that we will study in this paper) have been the **Morlet** wavelet (a Gaussian modulated by a complex exponential) [7,8], **Daubechies** wavelets (compact support wavelets) and Wavelet of the **Mexican Hat** (the second derivative of a Gaussian).

Morlet: $$\psi(x) = \frac{1}{\sqrt{2\pi}} e^{-\frac{x^2}{2}} e^{-i\omega_o x} \qquad (8)$$

Mexican Hat: $$\psi(x) = \frac{2}{\sqrt{3}} \pi^{-\frac{1}{4}} (1 - x^2) e^{-\frac{x^2}{2}} \qquad (9)$$

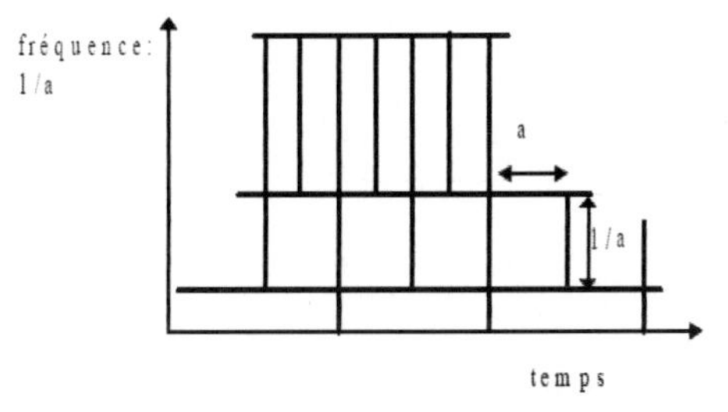

Figure 3. Time-frequency paving for the discrete wavelet transform.

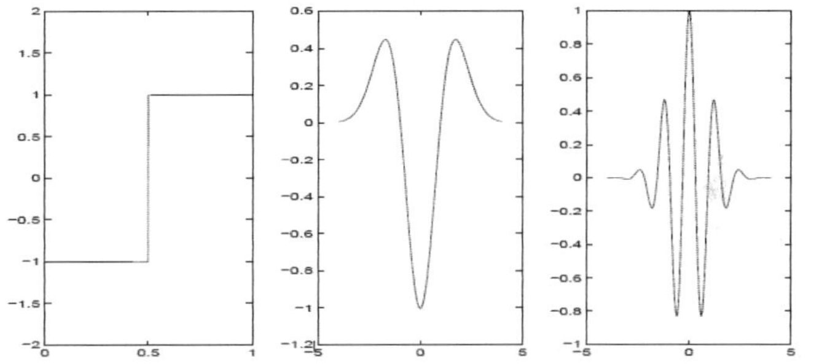

Figure 4. Different types of wavelets: Haar wavelet, Mexican hat wavelet, Morlet wavelet.

4.2. Principle of the Haar wavelet

The Haar wavelet is considered the first known wavelet. It is a piecewise constant function that will be denoted by H(x), and defined on the interval [0; 1] (or sometimes on [-1/2; 1/2]) which makes it the simplest wave to understand and implement moreover, its support is compact; it is well located in space. On the other hand, it has only one time zero and is discontinuous (equation 10 and figure 5):

$$H(x) = \begin{cases} 0 \ si & x < 0 \\ 1 \ si & 0 \leq x < \frac{1}{2} \\ -1 \ si & ½ \leq x < 1 \\ 0 \ si & 1 \leq x \end{cases} \qquad (10)$$

This wavelet is very simple and is therefore easy to implement algorithmically [8]. Moreover, its support is compact; it is well located in space. On the other hand, it has only one time zero, and is discontinuous.

4.3. Algorithm for edge detection by the Haar wavelet transform

Wavelet transform algorithm we used is based on a 5 x 5 sliding window with a pitch of 2 pixels in the horizontal and vertical directions. The calculations are shown in equation (11), where d_0 and d_{01} are the outputs for each column of five pixels (p_{11} to p_{15}). Thus, for five columns [9], one obtains 5 d_0 and 5 d_{01}. Then, the two columns of d_0 and d_{01} are computed again using the same equations.

$$d_0 = (-1 \ 2 - 1) * \begin{pmatrix} p_{11} \\ p_{12} \\ p_{13} \end{pmatrix}; \quad d_1 = (-1 \ 2 - 1) * \begin{pmatrix} p_{13} \\ p_{14} \\ p_{15} \end{pmatrix}; \quad d_{01} = (d_0 + d_1) > 3 \qquad (11)$$

IOP Conf. Series: Journal of Physics: Conf. Series **870** (2017) 012016 doi:10.1088/1742-6596/870/1/012016

Therefore, each 5 × 5 window generates 4 output pixels, as is a sliding window in a step of 2 pixels at a time in both horizontal and vertical directions; we always get an output pixel per input pixel on average.

In the analysis phase of the two-dimensional wavelet transform (see Figure 6), each row of the input image is filtered separately by H and G [low-pass (H) and high-pass (G)]. The resulting pair of line-transformed images is also filtered in the column direction, giving four sub-band images at [J = 1]. The three images of "details" or simply called LH (low-high), HL (high-low) and HH (high-high) correspond to specific bands, not superimposed in the frequency domain. The "smooth" or LL (low-low) S_{LL} component is a low-pass filtered version of the original image, and is transmitted to the next level for further decomposition of the sub-band.

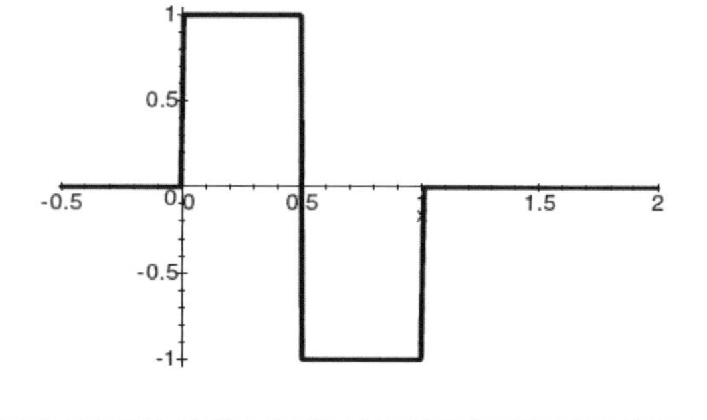

Figure 5. The analysis mother wave of Haar

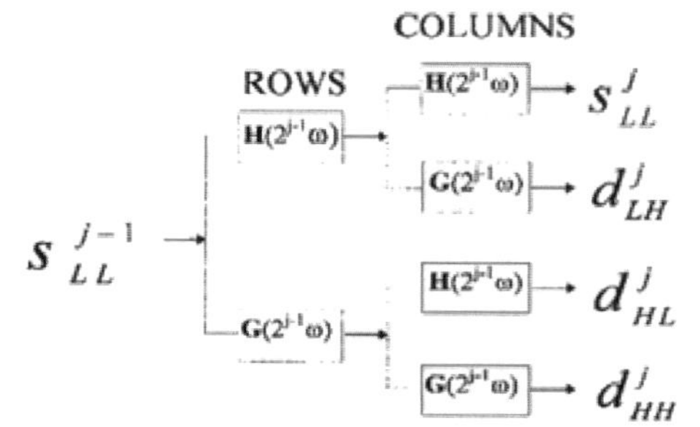

Figure 6. Wavelet Transform Filter Implementation

4.4. Results of applications of the Haar DWT algorithms on a digital image

In this section, we present the results of the application of Haar's DWT contour detection method [10]. The table shows the application on different direction of the image, the horizontal detection of the contours, the vertical detection, and the diagonal detection.

Table 1. Application of Haar's DWT for different directions

I	 Original digital image	 Gray level image
II	 Application of Horizontal Haar DWT	 Application of Vertical Haar DWT
III	 Application of diagonal Haar DWT	 Application of operator Canny

IOP Conf. Series: Journal of Physics: Conf. Series **870** (2017) 012016 doi:10.1088/1742-6596/870/1/012016

From the results obtained, we can conclude that the detection of the contours by DWT of Haar gives results more precise and closer to the results obtained by the classical operator of Canny, especially in the direction of the contours.

5. Real-time image processing equipment

The processing of the image in real time requires a high computing power. For example, image standard.jpg with a dimension of 640 * 480 approximately 0.25 mega pixels per image, with a calculated power of 100 frames per second for a 25 MHz processor. The size of the image may be larger; the amount of processing required per pixel depends on the image processing algorithm.

Haar's DWT contour detection algorithm is a combination of over ten steps, horizontal detection, vertical and diagonal detection, and their convolution product that scans all pixels.

As high-resolution images become more frequent, processing requirements will increase, high-resolution images of standards generally have ten times more pixels per image, the calculation load is approximately ten times higher, and they require more DSP or a single DSP, very expensive high-end. In this scenario, FPGAs offer real-time alternative platform image processing. FPGA effectively supports high levels of parallel processing of data flow structures, which are important for the effective implementation of image processing algorithms.

6. Hardware implementation of the algorithm on a Xilinx FPGA circuit

Technical For the implementation of our algorithm, we used the Xilinx nexys3 platform (a board that consists of a Spartan-6 FPGA circuit plus I / O interface) and VGA monitor to display the results, the algorithm was developed on the Xilinx ISE interface and the various blocks are programmed in VHDL.

6.1 *Block of "Memory" Program*

The program block "Memory" that exists in the IP part of the Xilinx software is configured according to the size of the image, a square image for our example of 40.000 pixel which will store the image and each memory cell containing 8 bits in binary value from 0 to 255 (the encoded value for each RGB pixel).

6.2 Block of "Memory Read" program

Since the image is not necessarily the same size as the screen, the screen size and 640*480 and the image size 200*200. The program is necessary for the correct positioning of the image on the screen, so this is the reading part of the pixels stored in the memory.

6.3 Block program "VGA screen"

The VGA monitor is controlled by five 10-bit coded signals: red (3-bit), green (3-bit), blue (2-bit), horizontal (1-bit) synchronization and Vertical synchronization (1-bit). The three color signals, collectively known as the RGB signal, are used to control the color of a pixel at a location on the screen. In order to produce other colors, each color analog signal must be supplied with a voltage between 0.7 and 1.0 volts to vary the color intensities.

6.4 Block program "Synchronization"

The program allows you to synchronize the scanning of the pixels on the screen one horizontally and the other vertically. The signals are used to control the timing of the scanning speed. The horizontal synchronizing signal determines the scanning time of a line, while the vertical synchronizing signal determines the scanning time of the entire screen. By manipulating these signals, images are formed on the monitor screen.

IOP Conf. Series: Journal of Physics: Conf. Series **870** (2017) 012016 doi:10.1088/1742-6596/870/1/012016

6.5 Block program "algorithm"

This part of the program presents the contour detection algorithm used in this paper (the principle is detailed in section 4).

After the conversion by applying the discrete wavelet transform, which is based on the FFT with a sliding window using the Haar wavelet, the pixels are always less than the total number of pixels [11].

The upper limit ensures that operations according to the FIFO list can be completed during the real-time processing period of a frame; FIFOs are implemented using multiple built-in M4K memory blocks, each 4 kbps in size. The image of the edge detection result is stored in the RAM blocks embedded in the Spartan-6 device. These large integrated RAM blocks can be used to temporarily store the edge image when 2 bits are allocated to each pixel location.

The following diagram shows the different blocks implemented on the software 'Xilinx ISE' [12,13]; software for programming the Xilinx FPGA circuits (figure 7).

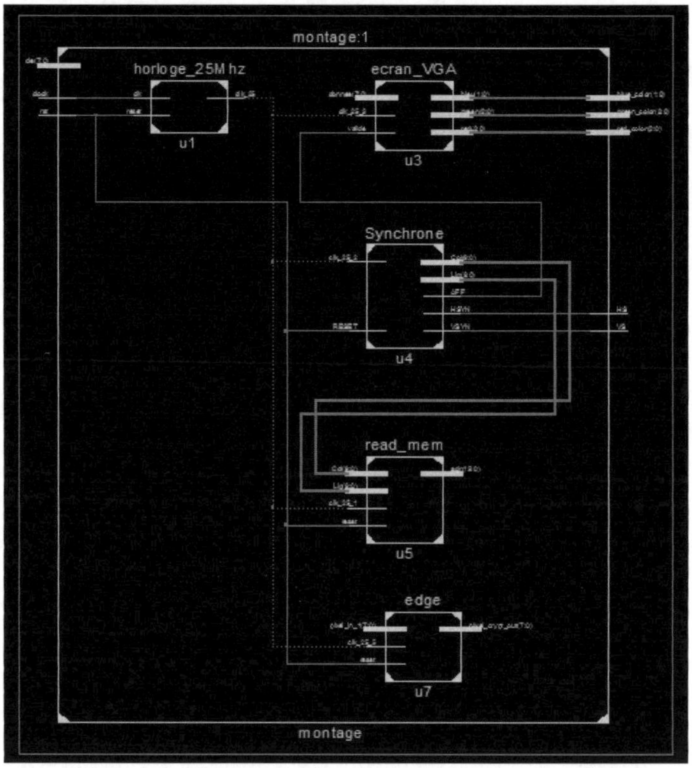

Figure 7. The blocks implemented on Xilinx ISE

7. Results and discussion

In Figure 7 We used the famous image in the field of image processing lenna.jpg for testing with a resolution of 200x200 pixels. The module is fully pipelined where a resulting pixel is computed at each clock cycle. With this rhythm and this clock frequency, it is possible to process more than 400 images at 200x200 pixels of resolutions.

Preliminary results show the FPGA design clocked at 100 MHz Spartan-6 targeting (figure 8).

The design is scalable to handle high-resolution images while maintaining the clock frequency used to process standard resolution video at 30 frames per second, as well as high-speed computer vision applications that require more than 80 images per second.

The use of high-resolution images requires a more powerful processor that exceeds the Spartan range, speaking of the Virtex range [14], high performance at frequencies above 500MHz. It is also possible to increase the frequency above 100MHz by introducing more pipeline stages to the detriment of an increased and optimal use of resources.

Figure 8. The image results before and after treatment.

8. Conclusions

This paper proposed the implementation of edge detection algorithms on FPGA architecture. The method presented is based on edge detection using the Haar discrete wavelet transform. The image is considered to be a 2D signal, we can pass in the frequency domain and in this case an edge can be seen as representing of the high frequencies of the signal.

The results obtained show the clear presence of all the edges that exist in the images (horizontal, vertical, and diagonal edges) [15,16]. The use of a spartan6 FPGA board demonstrates its ability to handle high-resolution VGA output images while maintaining the clock frequency used (100MHz) to process high-resolution image sequences at 30 frames per second as well as high-speed computer vision applications, which require more than 80 frames per second.

From the results obtained, we can conclude that the detection of the contours by DWT of Haar gives results more precise and closer to the results obtained by the classical operator of Canny, especially in the direction of the contours. For the improvement and performance of this work for the real-time processing of images, we propose as a prospect the use of a CMOS camera for the acquisition of real images, this camera will be connected to the FPGA card spartan6 Using a Pmod port connector, the results of the tightly processing images display permanently on a VGA screen.

9. References

[1]. T. Lindeberg, "*Edge Detection and Ridge detection with automatic scale selection*", International Journal of Computer Vision, Vol. **30**, No. 2, 117-154, 1998.

[2]. J. Malik, S. Belongie, T. Leung and J. Shi, "*Contour and texture analysis for image segmentation*", IJCV, vol.**43**, no.1, 2001.

[3]. Daubechies, Ten Lectures on Wavelets. Philadelphia, PA: SIAM, 1992.

[4]. Cohen, I. Daubechies, and J. C. Feaurean, "*Biorthogonal bases of compactly supported wavelets*", Commun. Pure Appl. Math., vol. **45**, 1990.

[5]. S. Mallat, W. Hwang, "*Singularity detection with wavelets*", IEEE Trans. Info. Theory, vol.**38**, pp. 617-643, 1992.

[6]. S. Mallat and S. Zhong, "*Characterization of signals from multiscale edges*", IEEE Trans. Pattern Anal. Machine Intell. vol. **14**, pp. 710–732, July 1992.

[7]. S. Mallat, "*A Wavelet Tour of Signal Processing. Orlando*", FL: Academic, 1999.
[8]. Grossman and J. Morlet, "*Decomposition of hardy functions into square integrable wavelets of constant shape",* SIAM J. Math Anal. vol. **15**, pp. 723-736, 1984.
[9]. M. Angelopoulou, K. Masselos, P. Cheung, and Y. Andreopoulos, "*A Comparison of 2-D Discrete Wavelet Transform Computation Schedules on FPGAs*", in IEEE International Conference on Field Programmable Technology, 2006, pp. 181-188.
[10]. J. Jyotheswar and S. Mahapatra "*Efficient FPGA implementation of DWT and modified SPIHT for lossless image compression*" Journal of Systems Architecture **53**, 369-378, 19 January 2007.
[11]. M.E. Angelopoulou, P.Y.K. Cheung, K. Masselos and Y. Andreopoulous "*Implementation and comparison of 5/3 Lifring 2D Discrete Wavelet Transform Computation Schedules on FPGAs*" Journal of Signal Systems **51**, 3-21, 2008.
[12]. http://store.digilentinc.com/fpga-programmable-log
[13]. http://www.xilinx.com/products/design-tools/ise-design-suite.html
[14]. S. Nazari, M. Amiri, K. Faez, and M. Amiri, "*Multiplier-less digital implementation of neuron–astrocyte signaling on FPGA*". Neurocomputing, 2015.
[15]. I. Bouganssa, M. Sbihi and M. Zaim "*Implementation of Edge Detection Digital Image Algorithm on a FPGA*" MATEC Web of Conferences. Vol. **75**, No 03003 Sept 2016.
[16]. I. Bouganssa, M. Sbihi and M. Zaim "*Implementation on a FPGA of Edge Detection Algorithm in Medical Image and Tumors Characterization*" DOI: **10**.1109/ICMCS.2016. 7905655. IEEE.Xplore 2016.

IOP Conf. Series: Journal of Physics: Conf. Series **870** (2017) 012017 doi:10.1088/1742-6596/870/1/012017

Piezoelectric actuators in the active vibration control system of journal bearings

J Tůma[1], J Šimek[2], M Mahdal[1], M Pawlenka[1], and R Wagnerova[1]

[1] Faculty of Mech Engineering, VSB – Technical University of Ostrava, 17. listopadu 15, CZ 708 33, Ostrava, Czech Republic
[2]TECHLAB Ltd., Prague, Sokolovská 207, CZ 190 00, Praha 9, Czech Republic

jiri.tuma@vsb.cz

Abstract. The advantage of journal hydrodynamic bearings is high radial load capacity and operation at high speeds. The disadvantage is the excitation of vibrations, called an oil whirl, after crossing a certain threshold of the rotational speed. The mentioned vibrations can be suppressed using the system of the active vibration control with piezoactuators which move the bearing bushing. The motion of the bearing bushing is controlled by a feedback controller, which responds to the change in position of the bearing journal which is sensed by a pair of capacitive sensors. Two stacked linear piezoactuators are used to actuate the position of the bearing journal. This new bearing enables not only to damp vibrations but also serves to maintain the desired bearing journal position with an accuracy of micrometers. The paper will focus on the effect of active vibration control on the performance characteristics of the journal bearing.

1. Introduction

The advantage of journal hydrodynamic bearings (alternatively called sleeve bearings or plain bearings for radial load) is high radial load capacity and operation at high speeds. The disadvantage is the excitation of shaft vibrations, called an oil whirl, after crossing a certain threshold of rotational speed which is dependent on the radial bearing clearance and the viscosity of lubricating oil. Once an unstable motion of the bearing journal occurs, the machine starts to vibrate, and its operating speed cannot be increased. Demonstration of the rotor instability during run-up is shown in figure 1.

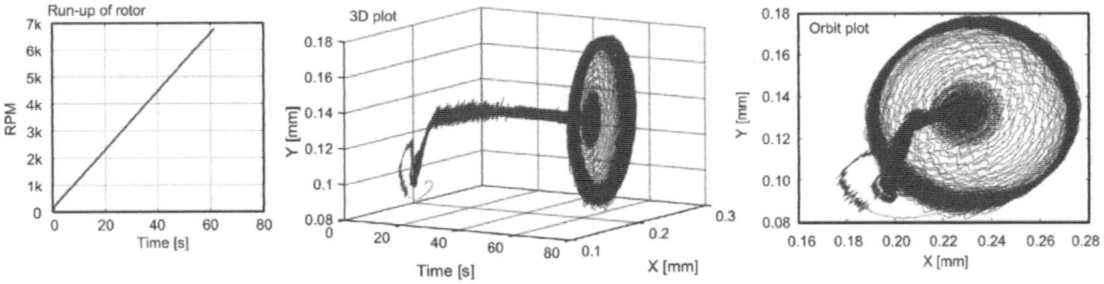

Figure 1. Journal hydrodynamic bearings instability due to the oil film.

The rotational speed of the rotor increases linearly with time at the slew rate of 7k rpm per minute. Journal equilibrium position is achieved after 20 seconds from the start-up. The bearing journal is moved

IOP Conf. Series: Journal of Physics: Conf. Series **870** (2017) 012017 doi:10.1088/1742-6596/870/1/012017

to the side while lifting up. A passive way how to suppress vibrations consists in adjusting the shape of the bearing bushing, such as lemon or elliptical bore of the bushing, a groove or tilting pads. Even though there are many solutions based on mentioned passive improvements, the approach to preventing the journal bearing instability is based on the use of the active vibration control (AVC) in the paper. To study possibilities of affecting rotor behaviour by controlled movement of bearing bushings, a test rig was designed, manufactured and assembled. The development work resulted in the design of the active bearings control system, which became the first functional bearing prototype in the known up to now [1]. In the introduction, it should be emphasized that research of active vibration control was aimed at rigid rotors and the standard design of cylindrical journal bearings, where the journal displacement is measured at the closest position to the bearing bushing.

1.1. Active vibration control

Vibrations of the shaft can be suppressed using the system for an active vibration damping with piezoelectric actuators to move the floating bearing bushing in two directions. The motion of the bearing bushing is controlled by the electronic controller, which responds to the change in position of the bearing journal with respect to the bearing housing. The mechanical arrangement of the actively controlled bearing is shown in figure 2. Two stacked linear piezoactuators are used to actuate the position of the bearing journal via the position of the bearing bushing as is shown in figure 3. The position of the journal or shaft is sensed by a pair of capacitive sensors with accuracy better than 1 micrometer. It works with a cylindrical bushing which did not require special technology of production and assembly. This new bearing enables not only to damp vibrations, but also serves to maintain the desired bearing journal position with an accuracy of micrometers.

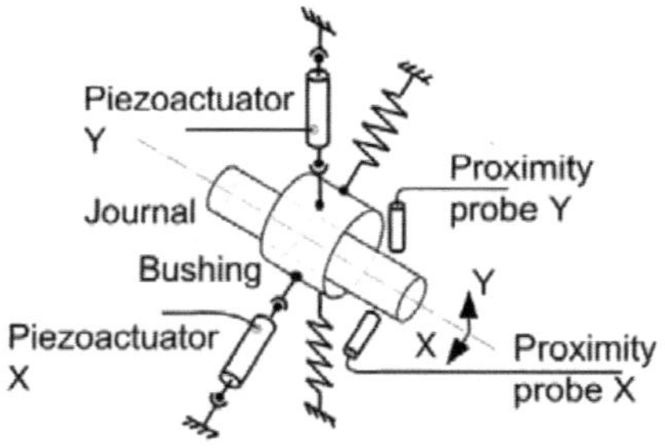

Figure 2. Actively controlled journal bearing.

Figure 3. Test bench for proof of concept.

1.2. Past developments

Many authors pay attention to the active control of journal bearings with the use of active magnetic bearings (AMB) [2] and giant magnetostrictive material (GMM) [3]. Although the authors of the paper on using GMM state that experiments can be carried out up to 1700 rpm, they publish only measurements at 350 rpm. The instability due to the oil film is a problem of high-speed rotors. With utmost probability, instability could not arise at such low rotational speed as 350 rpm. Therefore, the active vibration control was not aimed at eliminating instability of journal bearings, but only at positioning the shaft axis. Piezoactuators as a tool to control of rotating machines have been intensively investigated in the literature since the end of 1980's [4]. The linear piezoactuators can create a large force in a very small track. The advantage of journal bearings with piezoactuators is that the bearing bushing mounting stiffness remains unchanged in the case of a loss of electric power supply comparing AMB.

1.3. Test stand

The performance of the actively controlled bearing was tested on the test bench with the span of bearing pedestals of 200 mm. The bearing diameter is 30 mm and the length-to-diameter ratio is equal to about 0.77. The radial clearance of the journal is 55 micrometers. The instability of the oil whirl type starts at 2k rpm. The active vibration control extends stable operating rotational speed range up to 12k rpm. An inductive motor of 400 Hz drives the rotor and therefore the maximum rotational speed is 23k rpm. Input for the oil inlet is in the horizontal plane of symmetry of the bushing. The position of the journal is measured by a pair of the proximity probes which are capacitive sensors originated from the Micro-epsilon Company. The sensors are of the capaNCDT CS05 type with a measurement range of 0.5 mm. An advantage of the capacitive sensors is that it is not necessary to ground the shaft.

1.4. Piezoactuators

As was stated before the bearing bushings are actuated using the piezoactuators oriented in vertical and horizontal directions and fastened to the rig frame. The preloaded open-loop piezoactuators are of the P-844.60 type, the product of the Physik Instrumente Company. The piezoactuator require a low voltage amplifier with the 120 V peak value (LVPZT). The piezoactuator travel range is up to 90 micrometers, the pushing force is up to 3 kN and the pulling force is only up to 7 kN. The piezoactuator force balances the force effect of the oil film. Virtual motion of the unloaded piezoactuator is proportional to its control supply voltage $u_x{*} = kV$. The resulting motion u_x of the bearing bushing also depends on the load force as shown in the working graph of figure 4.

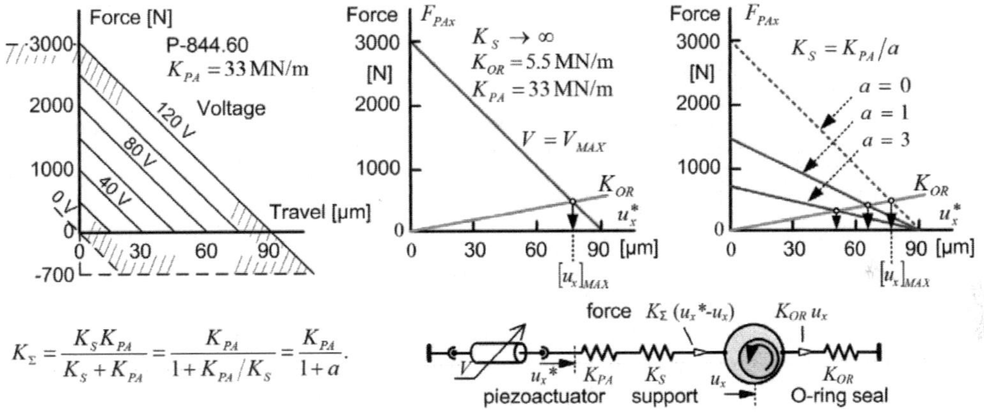

$$K_\Sigma = \frac{K_S K_{PA}}{K_S + K_{PA}} = \frac{K_{PA}}{1 + K_{PA}/K_S} = \frac{K_{PA}}{1+a}.$$

Figure 4. Mechanical part of the control loop

2. Control system and its effect on bearing operation

n electronic feedback is created for each of the two directions of the movement of the bearing journal in the bearing bushing. The control system is thus composed of two independent loops, each of which has its own controller. Both the controllers are of the proportional type. Although adding a derivative component improves the dynamic properties of the control loop, the corruption by noise of the proximity probes signal is the reason, for which the derivative feedback was not used. The controllers were created as a digital in the signal processor of the dSpace type. The input and output voltage of the DC amplifier is also shown in figure 5. The sampling frequency is chosen equal to 5 kHz.

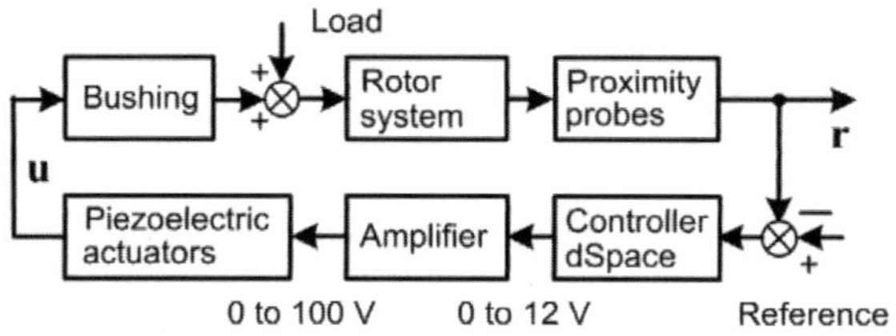

Figure 5. Active vibration control system.

2.1. Increasing the stable operating range of the bearing

The journal bearing has been tested at an increase in rotational speed until it becomes unstable and the journal motion is limited only by the bushing walls. The summary results are shown in figure 6. The dependence of the speed on time is shown in the left panel of figure 6, and the other three panels of this figure on the right show the dependence of the position of the bearing journal on time. An unstable motion of the journal is shown in the first of the upper mentioned three panels; the motion during active vibration control is on the second and third panel. The motion which is parametrically damped is the rightmost. The gain of the open loop of the control circuit was 35.

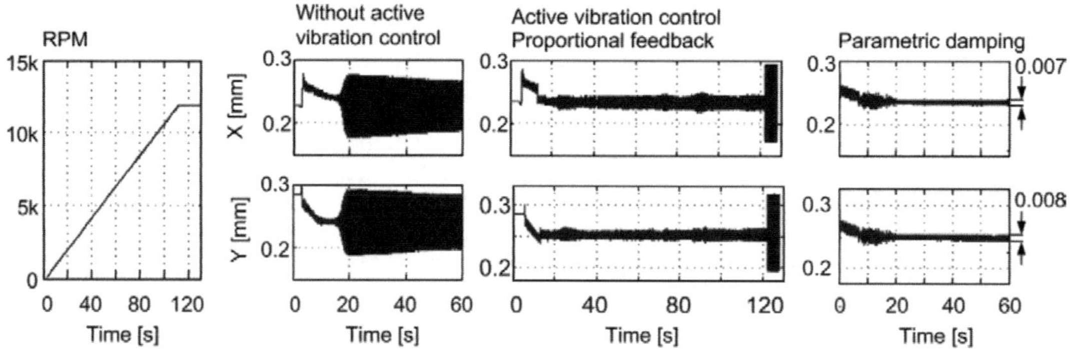

Figure 6. Summary results of the active vibration control operation.

The amplitude of the residual oscillation of the journal does not exceed 8 micrometers. Precision ball bearings (Deep groove ball bearings) which are offered by SKF have a radial internal clearance C2 to a diameter of 30 mm in the range from 1 to 11 micrometers. The maximum rotational speed of the 206-SFFC bearing type is only 7.5k to 13k rpm.

2.2. Reducing friction losses of the journal bearing

Power losses in the journal bearings were estimated from the electric power which is consumed by frequency converter and motor. Dependence of electrical power upon the rotational speed of the motor was measured with and without active control as it is shown in figure 7. The basic power consumption of the motor and frequency convertor was measured with the disconnected clutch between the motor and rotor; it means that the bearings were inoperative. The friction loss of a pair of bearings at 7k rpm is 66 W in an unstable operation, and if the active vibration control is ON, then the friction loss is of only 48 W. The active vibration control reduces the friction losses of journal bearings by 27 %. As a lubricant, the hydraulic oil of the OL-P03 type (VG 10 grade, kinematic viscosity 2.5 to 4 mm2/s at 40 °C) was used. All tests were undertaken at ambient temperature about 20 °C.

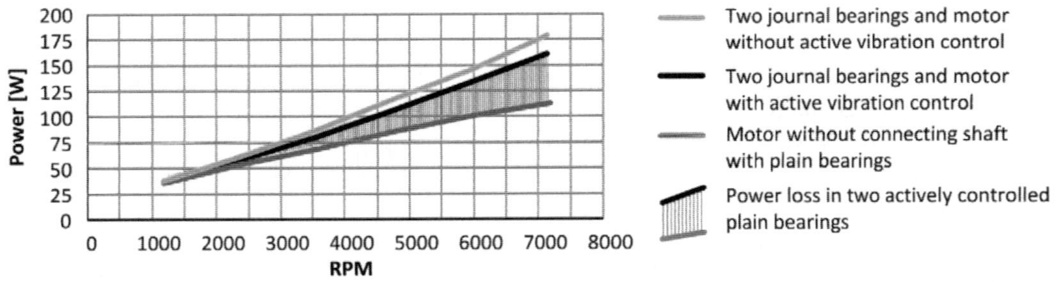

Figure 7. The electric power consumed by the frequency converter, motor, and bearings.

2.3. Increasing the radial stiffness of the journal bearing

The bearing bushing is suspended on a pair of piezoactuators and the bearing journal is supported by an oil wedge. According to the parameters in figure 4, the piezoactuator stiffness is of 33 MN/m. The stiffness of the O-ring seal is 5.5 MN/m. Force is transmitted to the bearing journal through the oil film. Based on the simulations it can be estimated that the direct stiffness of the oil film in the neighbourhood of the central position within the bushing bore is of 185 kN/m and the quadrature stiffness is still an order of magnitude larger and increases proportionally to the rotational speed. This stiffness increases by many orders of magnitude if the journal is approaching the bearing bushing wall. The stiffness of the journal support is defined first of all by the stiffness of the oil film. A steady-state error in a non-controlled bearing originates in a radial load which can be considered as a disturbance. A proportional controller which governs the system of journal bearings in the closed-loop with the open-loop gain of 35 reduces the steady-state error 36 times what results in the increase of the oil wedge stiffness 35 times compared to the design without feedback. Allowable forces, however, are limited by the load capacity of the piezoactuators. Notice that on the market there are piezoactuators enabling to generate forces up to 20 kN. The stiffness of the precision rolling bearings ranges from 100 to 200 kN/m, regardless of the load, while the stiffness of hydrodynamic bearings in neighbourhood of the central position (low load) is of the order of several kN/m. However with the active control, the stiffness can increase as much as 36-times, i.e. it can achieve values around of 100 kN/m, which is comparable to that of the precision ball bearing.

3. Conclusions

The advantage of journal hydrodynamic bearings is high radial load capacity and operation at high speeds. The disadvantage is the excitation of shaft vibrations, called an oil whirl, after crossing a certain threshold rotational speed which is dependent on the radial bearing clearance and the viscosity of lubricating oil. that provides these benefits as follow

- increasing the stable operating range of the bearing;
- reducing the friction losses of the journal bearing;
- increasing the radial stiffness of the journal bearing.

The cylindrical bushing does not require an exclusive technology of production and assembly. Electronic feedback improves the properties of the plain bearing and reduces the cost of production.

References

[1] Tůma J., Šimek J., Škuta J. and Los J. (2019): Active vibrations control of journal bearings with the use of piezoactuators. *Mechanical Systems and Signal Processing*, Vol. **36**, 2013, p. 618-629.

[2] Fürst, S. and Ulbrich, H. (1988): An Active Support System for Rotors with Oil-Film Bearings, In: *Proceedings of IMechE, Serie C, 1988*, pp. 61-68, paper 261/88.

[3] Lau, H.Y., Liu, K. P., Wang, W. and Wong, P. L. (2009): Feasibility of Using Giant magnetostrictive material (GMM) Based Actuators in Active Control of Journal Bearing System,

In: *Proceedings of the World Congress on Engineering 2009*, Vol II, WCE 2009, July 1 - 3, 2009, London, U.K.

[4] Palazzolo, A.B., Lin, R. R., Alexande, R. M., Kascak, A. F. and Montague, G. (1991): Test and Theory of Piezoactuators - Active Vibration Control of Rotating Machinery, *ASME Trans. Journal of Vibration and Accoustics*, 1991, 113(2) 167-175.

Acknowledgments

Authors wishing to acknowledge the financial support from the Czech Grant Agency, namely by the Project No. P101/12/2520 “Active Vibration Damping of Rotor with the Use of Parametric Excitation of Journal Bearings”. Valuable support was also within the Project Pre-Seed Activity Technical University of Ostrava II – Engineering, the individual activity IA01 „Actively Controlled Bearing” No. CZ.1.05/1.3.00/14.0318 and within the Project No. SP2016/84 – „Modern Methods of Machines and Processes Control”.

IOP Conf. Series: Journal of Physics: Conf. Series **870** (2017) 012018 doi:10.1088/1742-6596/870/1/012018

Experimental research on durability of recycled aggregate concrete under freeze- thaw cycles

Yanqiu Cheng[1, a *], Xiaoyu Shang[2,b], and Youjia Zhang[2,c]

[1]Engineering Section, Northeast Dianli University, Jilin, China,132012
[2]School of Civil Engineering and Architecture, Northeast Dianli University, Jilin, 132012
[a]chengyanqiu_nedu@sina.com, [b]shangxujin2@163.com, [c]zhangyoujia@nedu.edu.cn

Abstract. The freeze-thaw durability of recycled aggregate concrete has significance for the concrete buildings in the cold region. In this paper, the rapid freezing and thawing cycles experience on recycle aggregate concrete was conducted to study on the effects of recycle aggregate amount, water-binder ratio and fly ash on freeze-thaw durability of recycle aggregate concrete. The results indicates that recycle aggregate amount makes the significant influence on the freeze-thaw durability. With the increase of recycled aggregates amount, the freeze-thaw resistance for recycled aggregate concrete decreases. Recycled aggregate concrete with lower water cement ratio demonstrates better performance of freeze-thaw durability. It is advised that the amount of fly ash is less than 30% for admixture of recycled aggregates in the cold region.

1. Introduction

Recycled aggregate concrete, as a kind of green sustainable building materials, can effectively alleviate the plight that a mass of construction waste is difficult to handle in cities. According to statistics, 3.5 billion tons of construction waste outputs every year in China, but the comprehensive utilization rate is less than 5%. About anti-freeze performance of the recycled concrete has been carried on by many scholars[1-6], but their research results presents inconsistent, and even opposite conclusions. Based on the fast freeze-thaw test, this paper will investigate the influence of admixture amount of recycled aggregates, water-binder ratio and fly ash content on the durability performance of recycled concrete, which can provide basis data and reference for the popularization and application of recycled concrete in cold areas.

2. Experimental program

2.1. Materials

P.O. 42.5 ordinary Portland cement was used. fly ash is chose by II grade fly ash. fine aggregate is the river sand, natural coarse aggregate is calcareous rock, efficient AE water reducer is used as the additive. Recycled coarse aggregate was prepared by waste concrete (original strength grade C30) that was crushed by the crusher shattered, then handled by 5 mm diameter of sieve screening process, meanwhile ensuring the aggregate controlled in 5-30 mm of diameter. The recycled coarse aggregate after screening was conducted by some program such as washing and drying to make it have excellent properties. Table 1 is the mix proportion of different categories of recycled aggregate concrete. to ensure the quality of recycled aggregate concrete specimens, the production program was conducted according to the Chinese rule DB11/T 803-2011. To ensure the quality of recycled concrete block, forced small mixer was used to

IOP Conf. Series: Journal of Physics: Conf. Series **870** (2017) 012018 doi:10.1088/1742-6596/870/1/012018

mix recycled aggregate concrete and recycled aggregate will be in the water-saturated state after prewetting processing. AE efficient water reducing agent, cement, aggregate, water and natural regeneration aggregate is successively put into the mixer. Mixed Materials are placed into the standard moulds after the testing recycled aggregate concrete workability.

Table 1. Mix proportion of specimens

Number	water-binder ratio	recycled aggregates amount	parameters of fly ash
A1	0.45	0	30%
A2	0.45	50%	30%
A3	0.45	100%	30%
B1	0.35	50%	30%
B2（A2）	0.45	50%	30%
B3	0.55	50%	30%
C1	0.45	50%	0
C2	0.45	50%	10%
C3（A2）	0.45	50%	30%
C4	0.45	50%	50%

2.2. Test measurement

In the light of designed mix proportion, the specimens (100mm×100mm×100mm and 100mm×100mm×400mm) will be conducted by mass measurement and dynamic elasticity modulus measurement respectively. Before freezing-thawing cycled experiment, the specimens need to be placed into the water(15℃~20℃) to reach the water-saturated state. Then, the fast freezing-thaw cycled experiment is conducted by the Chinese rule GBJ82-85. The duration of each freeze-thaw cycles is 4 hours. The specific procedure is shown inTable 2. Mass and dynamic elastic modulus is measured after every completed 25 times freeze-thaw cycle. It is a total of 300 freeze-thaw cycling test. The test will be stopped when the specimen with a 5% drop on mass or dynamic elastic modulus decreased to less than 60%.

Table 2. Procedure for freeze-thaw cycles

Type	Initial time	Final time	Temperature (℃)
Stage 1	0:00	0:10	15
Stage 2	0:10	0:40	15~-20
Stage 3	0:40	2:10	-20
Stage 4	2:10	2:30	20
Stage 5	2:30	4:00	20

3. Results and analysis

The different performance degradation rule of recycled aggregates concrete specimens after 300 times fast freeze-thaw test is shown. And the relative results with the consideration of recycled aggregates, water-binder ratio, fly ash admixture are presented in the following parts.

Different recycled aggregate parameter changes impact on the quality of concrete specimens after freezing and thawing cycle. It can be seen from the Fig 1, Fig.2 and Fig.3 that concrete mass loss after freeze-thaw cycle increases with the amount of recycled aggregates parameter rising. There are no more than 5% mass loss in A2 group after 300 times of freeze-thaw cycle. It is indicated that the recycled aggregate concrete in the terms of mass loss can be accepted. When amount of recycled aggregate reaches to 100%, the mass loss is so severe for A3 after 300 times freeze-thaw cycles. mass loss is not more than 3% for different water-binder ratio of recycled concrete specimens after 300 times fast freeze-thaw cycle which presents well freeze-thaw performance. . But it can be seen that water-binder ratio has a great influence on the quality loss. The smaller water-binder ratio of recycled aggregates concrete is and the smaller mass loss is. For the specimens of C3 group (30% fly ash amount), the mass loss is not more than 5%. When fly ash amount is 50% for C4 group, the mass loss is more than 5%. Thus, it is obvious that the no more than of amount of fly-ash for recycled aggregate concrete is advisable.

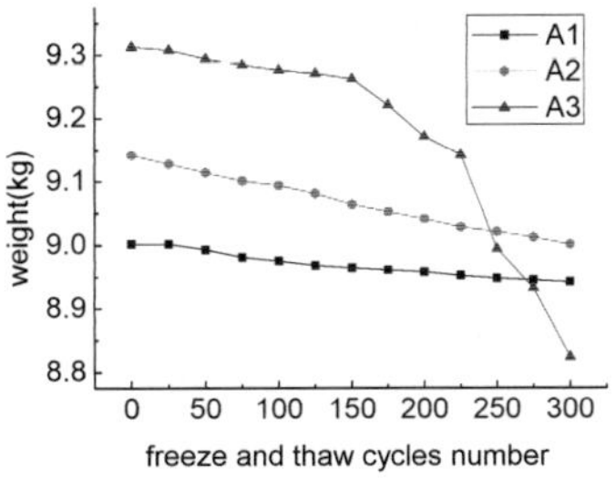

Fig.1Mass loss with diffrent recyled aggragates

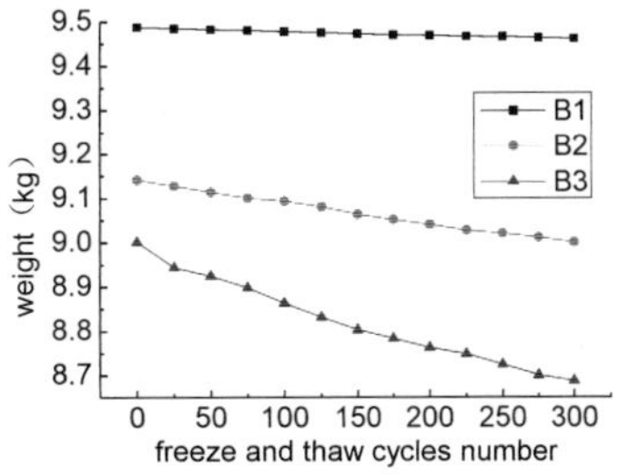

Fig.2 Mass loss with diffrent water-blind ratio

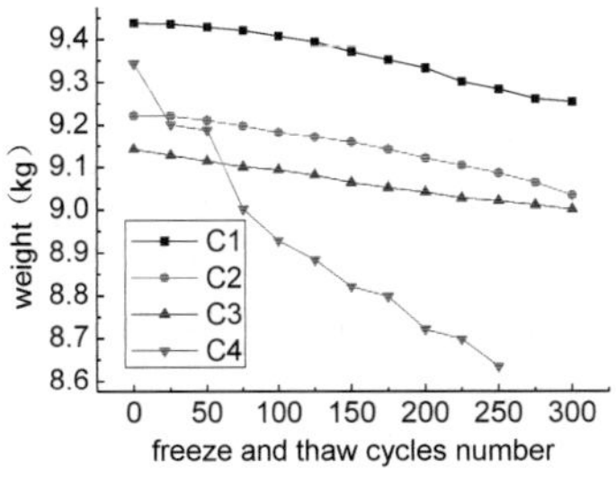

Fig.3 Mass loss with diffrent fly ash

The Fig.4 shows the effects of recycled aggregates, water-binder ratio and fly ash on dynamic elasticity modulus. It can be observed that these factors has different extents influence. When the amount of recycled aggregates is 50% for A2 group. The dynamic elasticity modulus still stays the level of 0.9. It demonstrates that freeze-thaw resistance of recycled aggregates concrete seems i=similar with ordinary concrete. With the increase of amount of recycled aggregates, the decline rate of dynamic elasticity modulus of recycled aggregates concrete for in the later stage of freeze-thaw cycles(200 times)is obviously faster when the amount of the recycled aggregates is 100% for A3 group, the relative dynamic elasticity modulus is lower 0.9. it shows that the freeze-thaw performance is weaker than the ordinary concrete. In order to obtain the well performance concrete, lower 50%the amount of recycled aggregate is advisable.

As Fig.5 shown, the recycled concrete after freeze-thaw cycle 300 times relative dynamic elastic modulus were above 0.9 as shown in Figure 5, antifreeze performance is still good. Water-binder ratio on dynamic elastic modulus of smaller, but still show some regularity, dynamic elastic modulus increases with the smaller of water/cement ratio. The relative dynamic elastic modulus of after 300 times freeze and thaw cycles keep above 0.9 as shown in Figure 5, which shows the well freeze and thaw performance for recycled aggregates concrete. the water-binder ratio has small effects on the dynamic elastic modulus, but there is obvious rule that dynamic elastic modulus increase with the water-binder ratio decrease.

In the Fig.6, when recycled concrete admixture amount of fly ash in more than 50% (C4) specimen, the freeze-thaw cycle elastic modulus decreased after 300 times larger, and even destruction of specimen. But other groups of specimens (C1, C2 and C3 and C4), the relative dynamic elastic modulus were above 0.9, shows that still has better performance of antifreeze. Experimental data show that in the process of 300 freeze-thaw cycles, not with fly ash of relative dynamic elastic modulus of recycled concrete specimens of lower speed no faster, with 10% and 30% fly ash of the recycled concrete specimens after freezing and thawing cycle 200 times, its relative dynamic elastic modulus decreased obviously accelerated, and fly ash admixture amount for 50% of C4 group of specimens, the relative dynamic elastic modulus after freezing and thawing circulation of 150, presents the obvious acceleration tendency.

IOP Conf. Series: Journal of Physics: Conf. Series **870** (2017) 012018 doi:10.1088/1742-6596/870/1/012018

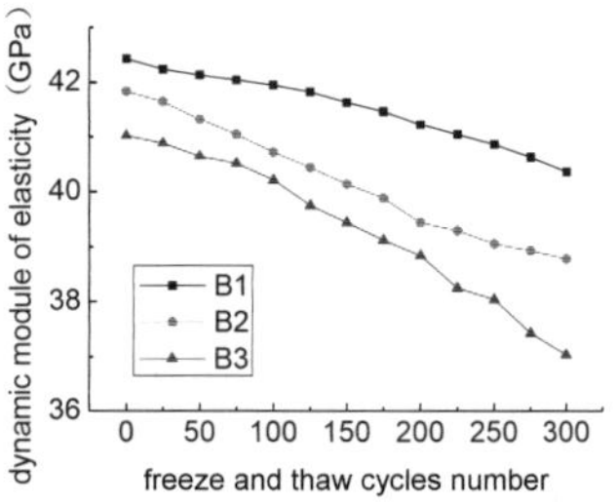

Fig.4 DEM with differrent recyled aggragates

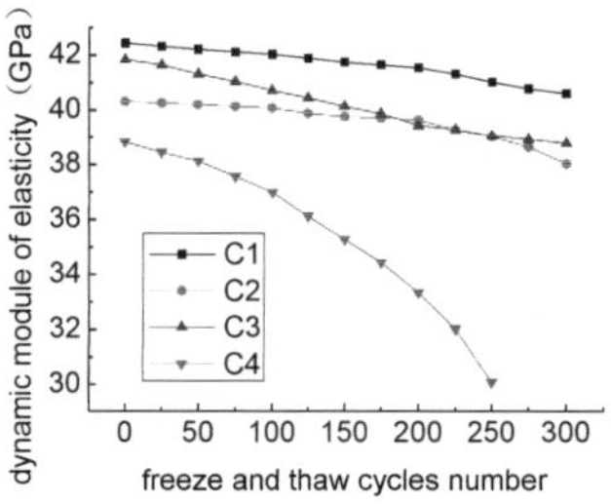

Fig.5 DEM with differrent water-blind ratio

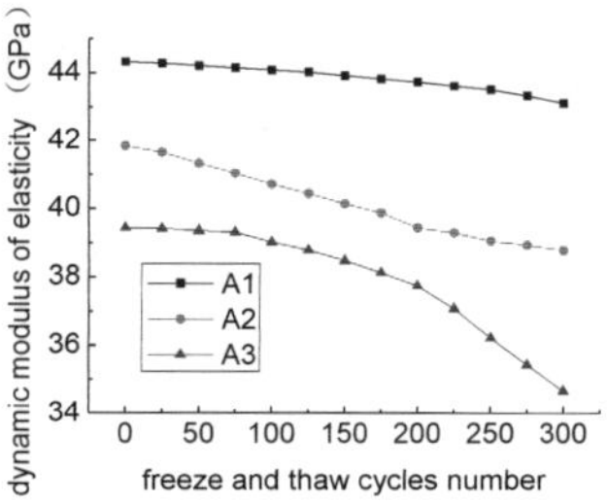

Fig.6 dem with differrent fly ash

4. Conclusion

The general rule of mass loss and dynamic elasticity modulus is obtained based on the 8 groups of recycled aggregates concrete specimens under freeze-thaw cycles test. The amount of recycled aggregates has an important role in the freeze-thaw performance of recycled aggregates concrete, which makes the declining tendency decreases with the enhancement of recycled aggregates . The concreted with 50% recycled aggregates shows the well performance in all the specimens. Water-binder ratio is also an essential factor on the freeze-thaw performance of recycled aggregates concrete. The specimens with lower water-blinder ratio presents the well freeze-thaw performance, whose mass loss increase and dynamic elasticity modulus decrease with the amount of fly ash enhancing. it is advised that the amount of fly ash is less than 30% for admixture of recycled aggregates in the cold region

References

[1] Levy S M, Helene P. Durability of recycled aggregates concrete: a safe way to sustainable development. Cement and Concrete Research. Vol. 34 (2004), p.1975-1980.

[2] Gokee A, Nagatakib, Saekic T and Hisadad M. Freezing and thaw resistance of air-entrained concrete incorporation recycled coarse aggregate: The role of air content in demolished concrete. Cement and Concrete research. Vol. 34(2004), P. 799-806.

[3] Li Qiuyi, Tian Li, Zhu Yaguang, Yang Xiangning, Zhu Chongji. Experimental study on durability of recycled concrete. Acta Scientiarum Naturalium Universitatis Sunyatseni. Vol.46(2007), P. 337-338.

[4] Chen Aimei, Zhang Qing, Wang Jing, Gai Zhanfang. Freez-thaw cycle test and a damage mechanics model for recycled concrete. Engineering Mechanics. Vol.26(2009), P.102-107.

[5] Cui Zhenglong, Ohaga Yoshiki, Kitatsuji Masa Humi, Tanaka ReiJi. Experimental research on freezing-thawing cycle of recycled aggregate concrete. Journal of building materials. Vol.10(2007), P. 532-537.

[6] Huo Yanhua, Mao Tiandian, Xiong Jingang. Experimental study on durability of recycled coarse aggregate concrete[J]. Sichuan Building Science. Vol.38(2012), P. 191-194.

IOP Conf. Series: Journal of Physics: Conf. Series **870** (2017) 012019 doi:10.1088/1742-6596/870/1/012019

Experimental Study on Abrasive Water Jet Machining of PZT Ceramic

Ajit Dhanawade[1], Ravi Upadhyai[1], Arunkumar Rouniyar[1] and Shailendra Kumar[1,a]

[1]Department of Mechanical Engineering, S V National Institute of Technology, Surat-395007, India

[a]skbudhwar@med.svnit.ac.in

Abstract. This paper presents research work involved in abrasive water jet machining of PZT ceramic material. Process parameters namely stand-off distance, water pressure and traverse rate are considered in the present study. Response surface methodology approach is used to design the experiments. Relative significance of process parameters and their influence on kerf properties are identified on the basis of analysis of variance. It is found that water pressure and traverse rate are most significant parameters followed by stand-off distance. On the basis of experimental analysis, regression models are developed to predict kerf taper and depth of cut. The models are developed with respect to significant parameters, interaction and quadratic terms. It is found that model predictions are in congruence with experimental results. Multi-response optimization of process parameters is also performed using desirability approach in order to minimize kerf taper and maximize depth of cut. Kerf wall features of machined surfaces are observed using scanning electron microscope. The findings of present study are useful to improve kerf properties in abrasive water jet machining of PZT ceramic materials.

1. Introduction

Lead Zirconate Titanate (PZT) is a metallic oxide based piezoelectric material. It has excellent piezoelectric properties including high piezoelectricity, Curie temperature, spontaneous polarization, sensitivity, dielectric constants, wear and corrosion resistance, thermal resistance, etc. [1]. It is widely used in ultrasonic cleaners, ultrasonic imaging, sensors, actuators, dielectrics for capacitors, buzzers, accelerometer, hydrophones, and ultrasonic motors, etc. [2]. Many of these applications demand intricate shapes, tolerances, and precise dimensions. However, machining of this material is difficult due to its brittle nature and temperature sensitivity. Traditional machining such as turning, milling, drilling, and diamond saw cutting has many difficulties including high tool wear due to high hardness and strength, friction and vibration, poor surface quality and breakage of material during machining [3]. Also, low material removal rate by traditional machining results in high processing cost. Abrasive water jet machining (AWJM) is one of non-traditional machining processes extensively used in the manufacturing industries for material processing. The unwanted material of work piece is removed by the combined action of high speed abrasive water jet (AWJ). AWJ impinges on the surface of the workpiece and removes the material by means of erosion. The major difficulty in AWJM of ceramics is improper kerf properties [4]. Kerf properties include kerf top width, kerf bottom width, kerf walls and kerf taper. The kerf of AWJ machined workpiece is characterised by wide kerf top and narrow kerf bottom [5]. Consequently it results in high kerf taper as depicted in Figure 1.

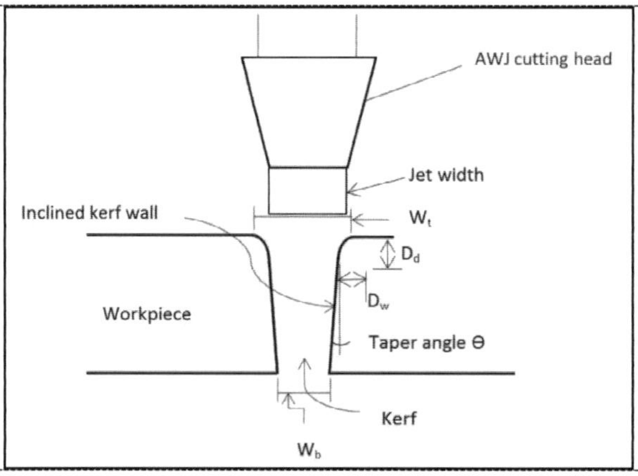

Figure 1. Kerf geometry in AWJM

Some researchers have applied their efforts to improve cutting performance of AWJM of ceramic materials. For example, Hocheng and Chang [6] experimentally studied mechanism of material removal, attainable depth of cut, kerf shape and surface roughness in AWJM of alumina oxide and silicon nitride ceramics. Chen et al. [7] investigated the influence of process parameters on the attainable depth of cut, depth of upper zone, kerf quality in AWJM of alumina ceramic. Momber et al. [8] investigated material removal process and specific removal energy in AWJM of refractory ceramics. A set of process parameters was optimized for depth of cut and MRR. Gudimetla et al. [9] studied material removal mechanism and kerf characteristics in AWJM of alumina ceramic. Wang and Guo [10] experimentally studied multi-pass cutting in AWJM of alumina ceramic. Effect of number of passes, traverse rate and traverse direction on kerf profile, depth of cut, and surface roughness was studied. Wang and Liu [11] developed predictive models to estimate kerf taper angle, depth of cut, and kerf top width in AWJM of alumina ceramic. The effect of process parameters on kerf quality of alumina ceramic by using kerf taper compensation technique is investigated by Shanmugam and Masood [12]. Srinivasu and Axinte [13] experimentally studied the effect of jet impingement angle on kerf top width in AWJM of silicon carbide ceramic. Annoni et al. [14] experimentally studied the influence of process parameters on kerf taper and surface roughness in AWJM of thin PZT ceramic sheets. Huang et al. [15] studied effect of AWJM process parameters on surface quality of polished aluminium nitride ceramic material. Ghosh et al. [16] experimentally studied the effect of process parameters on surface roughness in AWJM of silicon nitride ceramic. Dittrich et al. [17] investigated influence of process parameters on kerf characteristics, attainable depth of cut and material removal rate in AWJM of alumina ceramic. Chithirai et al. [18] experimentally studied influence of process parameters on surface quality of AWJ machined alumina ceramic.

From the literature review, it is clear that most of the researchers have studied AWJM of ceramic materials including alumina ceramic, silicon nitride ceramic and aluminium nitride ceramic. Researchers have focused on influence of process parameters namely pressure, traverse rate, stand-off distance, abrasive mass flow rate etc. on kerf properties, material removal rate, machinability, etc. Very few researchers have applied their efforts to study kerf properties of AWJ machined PZT ceramic. In the present study, AWJM of PZT ceramic is carried out to investigate effect of process parameters on kerf taper and depth of cut. Second order mathematical models are developed to predict kerf taper and attainable depth of cut on the basis of analysis of variance (ANOVA) results. Also optimization of process parameters is performed to minimize kerf taper and maximize depth of cut.

2. Experimental work

A computer controlled flying arm AWJ machine is used for experiments. Pressure is controlled by dial indicator on pump having maximum pressure of 220 MPa. Throughout the experiments garnet abrasives of mesh #80 size are used, delivered through gravity feed hopper to the nozzle. The nozzle diameter, orifice diameter, impingement angle and focusing length were kept constant at 0.76 mm, 0.25 mm, 90^0 and 70 mm respectively. The properties of workpiece material (PZT-5H) are given in Table 1. The thickness of workpiece material was 16 mm. This material is used further to manufacture ceramic matrix composite of around 14-15 mm thickness.

Table 1. Properties of workpiece material

Property		Value
Density (kg/m3)	ρ	7500
Elastic compliance	sE_{11}	15
(x 10^{-12} m^2/N)	sE_{33}	21
Dielectric constant	K^T_3	3250
Mechanical quality factor	Q_M	65
Electrical quality factor	Q_E	40
Curie Temp. (°C)	T_c	190

2.1 Experimental design

Central composite design (CCD) of response surface methodology (RSM) is used to design experiments. Experimental design includes 8 factorial runs, 6 axial runs and 6 centre points. Therefore, total 20 experiments are carried out according to the combinations suggested by Design Expert v10 software. Process parameters namely stand-off distance, water pressure and traverse rate are considered. Levels of process parameters as listed in Table 2 are selected on the basis of literature review, trial experiments and available machine setup range. Experiments are performed according to given parameter combinations [19].

Table 2. Machining Parameters and selected levels

Parameter	Level		
	-1	**0**	**1**
Stand-off distance (SOD) mm	2	3	4
Water pressure (WP) MPa	140	150	160
Traverse rate (TR) mm/min	310	320	330

2.2 Measurement of kerf taper and depth of cut

After experimentation kerf widths and depth of cut are measured by using vision measurement system (Model- Sipcon SDM-TRZ 5300). Further kerf taper angle is calculated by using equation (1).

$$\theta = \tan^{-1} (W_t - W_b)/2t \quad (1)$$

Where, W_t = Top kerf width (mm)
W_b= Bottom kerf width (mm)
t = Thickness (mm)

3. Predictive models

The regression models are developed to predict kerf taper and depth of cut.

3.1 Kerf taper

3.1.1 ANOVA. Table 3 gives the ANOVA results for kerf taper. The analysis is carried out at 95% confidence level. The model is significant with F-value of 185.48. Traverse rate and water pressure are found most significant parameters followed by stand-off distance. Interactions between SOD-WP and WP-TR are significant. Likewise, quadratic terms of SOD and TR are significant.

Table 3. ANOVA for kerf taper

Source	**SS**	**DoF**	**MS**	**F**	**p**
Model	3.46	7	0.49	185.48	< 0.0001
A-SOD	0.94	1	0.94	351.81	< 0.0001
B-WP	1.03	1	1.03	387.36	< 0.0001
C-TR	1.21	1	1.21	452.96	< 0.0001
AB	0.029	1	0.029	10.72	0.0067
BC	0.035	1	0.035	12.94	0.0037
A^2	0.15	1	0.15	56.09	< 0.0001
C^2	0.11	1	0.11	40.82	< 0.0001
Residual	0.032	12	2.667×10^{-003}		
Lack of Fit	0.020	7	2.925×10^{-003}	1.27	0.4103
Pure Error	0.012	5	2.305×10^{-003}		
Corrected Total	3.49	19			

The regression model is developed with respect to significant terms. The final regression equation for kerf taper in terms of coded factors is as given in equation (2).

$$\text{Kerf taper} = 91.481 + 0.686\ SOD - 0.218\ WP - 0.482\ TR - 5.976 \times 10^{-3}\ SOD\ WP + 6.568 \times 10^{-4}\ WP\ TR + 0.0753\ SOD^2 + 6.43 \times 10^{-4}\ TR^2 \quad (2)$$

The predicted and adjusted R^2 value of 0.9720 and 0.9855 reveals reasonable agreement. It is found that predicted results are in congruence with the experimental results as depicted in the Figure 2. The equation is useful to predict kerf taper.

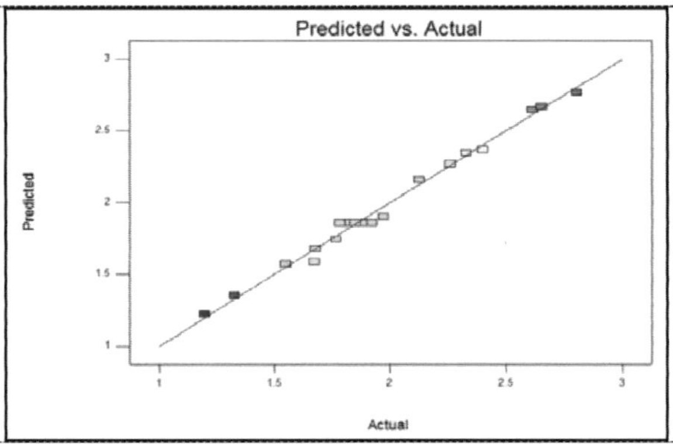

Figure 2. Predicted v/s actual values of kerf taper

3.1.2 Influence of process parameters on kerf taper. Figure 3 depicts effect of water pressure and stand-off distance on kerf taper. It is observed that increasing water pressure causes decrease in kerf taper. Increase in pressure results in increase in kinetic energy. The jet with increased kinetic energy cuts the material at bottom region which results in minimum kerf taper. Also it causes increase in effective diameter of the jet. It results in wide bottom width which consequently reduces kerf taper. It is observed that kerf taper increases with increase in stand-off distance. It is due to the fact that flaring of water jet occurs when it comes out of nozzle. It reduces cutting ability of jet and results in increased kerf taper.

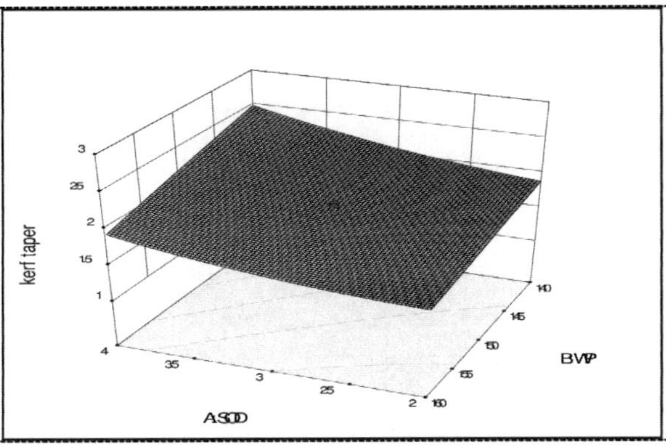

Figure 3. Effect of SOD and WP on kerf taper

Effect of water pressure and traverse rate on kerf taper is shown in Figure 4. It is observed that kerf taper increases with an increase in traverse rate. The reason behind this is high traverse rate reduces jet interaction on a given area of material. It causes material erosion by less number of abrasive particles and results in high kerf taper.

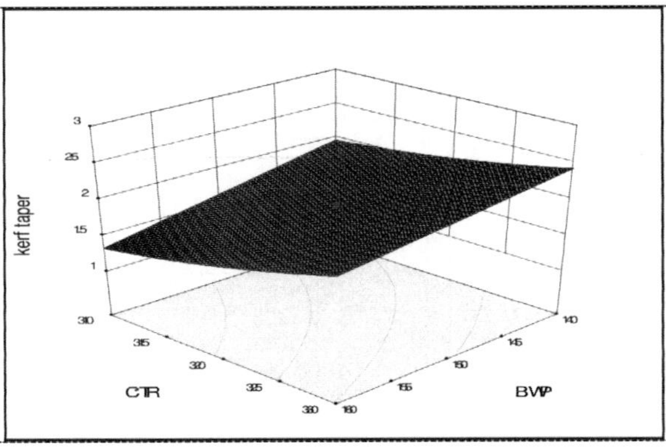

Figure 4. Effect of WP and TR on kerf taper

3.1.3 Kerf quality. Kerf geometries of four samples machined by setting parameters as given in Table 4 are shown in Figure 5. It is observed that kerf geometry of samples is characterized by round top edge. However, it is observed that roundness of top edge of sample 2 is more as compared to sample 1. This is due to an increase in stand-off distance which increases flaring of jet and results in rounding of top edge. It is also observed that rounding of top edge is more in sample 3 as compared to sample 4. This is due to increase in water pressure of sample 4. Kinetic energy increases with increase in pressure and minimizes stray abrasive particles, which results in sharp kerf edges. In non-through cuts,

IOP Conf. Series: Journal of Physics: Conf. Series **870** (2017) 012019 doi:10.1088/1742-6596/870/1/012019

ballooning effect is observed at bottom region. This effect occurs due to jet deflection, as there is no passage for water jet to flow.

Table 4. Parametric combinations of samples 1 to 4

Sample	SOD (mm)	WP (MPa)	TR (mm/min)
1	1	150	320
2	5	150	320
3	2	140	330
4	2	160	330

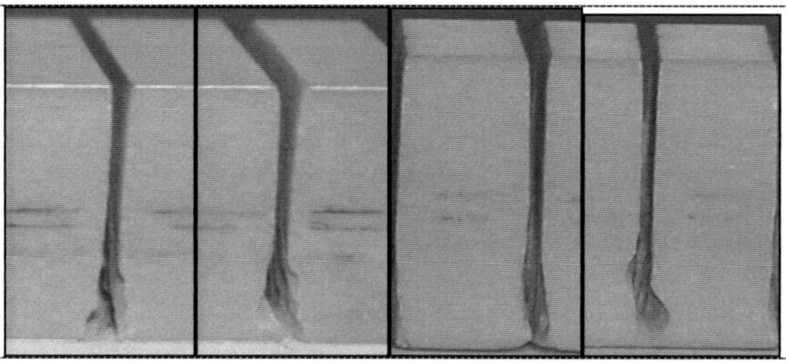

Figure 5. Kerf geometry of (a) Sample 1 (b) Sample 2 (c) Sample 3 (d) Sample 4

3.2 Depth of cut

3.2.1 ANOVA. Table 5 gives ANOVA results of depth of cut.

Table 5. ANOVA for depth of cut

Source	SS	DoF	MS	F	p
Model	7.86	4	1.97	103.01	< 0.0001
A-SOD	0.86	1	0.86	45.11	< 0.0001
B-WP	5.12	1	5.12	268.10	< 0.0001
C-TR	1.63	1	1.63	85.63	< 0.0001
B^2	0.25	1	0.25	13.20	0.0025
Residual	0.29	15	0.019		
Lack of Fit	0.17	10	0.017	0.75	0.6716
Pure Error	0.11	5	0.023		
Cor Total	8.15	19			

From ANOVA results, it is observed that model is significant with F value of 103.01. It is found that water pressure is most significant parameter followed by traverse rate and stand-off distance. Quadratic term of water pressure is also significant. No interaction term is significant.

From ANOVA results, a regression model is developed with respect to significant terms. The final regression in terms of actual factors is given in equation (3).

$$\text{Depth of cut} = -4.345507352941 - 0.2319375\ SOD + 0.34518345588235\ WP - 0.0319562\ TR - 0.00096213235294117\ WP^2 \quad (3)$$

The predicted R^2 value of 0.9260 is in reasonable agreement with adjusted R^2 value of 0.9555. The equation is useful to predict depth of cut. The predicted values of depth of cut are in congruence with experimental values as shown in Figure 6.

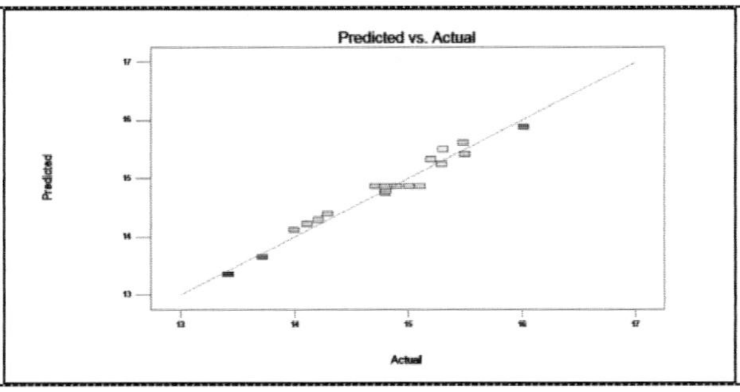

Figure 6. Predicted v/s actual values of depth of cut

3.2.2 Influence of process parameters on depth of cut. The variation in depth of cut according to experimental combinations is shown in Figure 7.

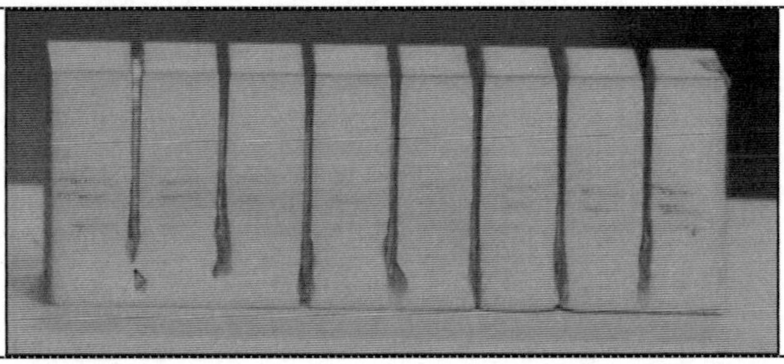

Figure 7. Variation in depth of cut on machines samples

Figure 8 depicts effect of water pressure and stand-off distance on depth of cut. Increase in water pressure results in increase in kinetic energy of jet. The jet with high kinetic energy penetrates deep in the material. It increases the depth of cut. Increase in stand-off distance results in flaring of AWJ. It reduces the cutting ability of jet. Therefore increase in stand-off distance reduces depth of cut. Figure 9 shows effect of traverse rate and water pressure on depth of cut. High traverse rate reduces number of abrasive particles impingement on material. It also results in less overlapping of machining action. Therefore increase in traverse rate results in decrease in depth of cut.

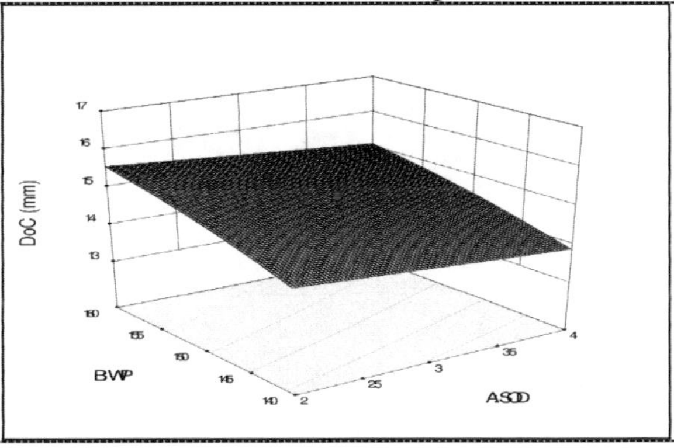

Figure 8. Effect of water pressure and stand-off distance

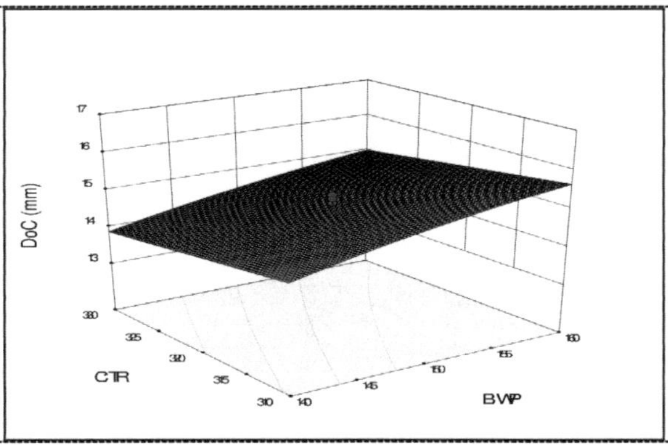

Figure 9. Effect of traverse rate and water pressure

3.2.3 Kerf wall features. Scanning electron microscope (SEM) is used to analyse kerf wall features of machined samples. SEM image of machined surface at three different regions i.e. top, middle and bottom are shown in Figure 10 [20]. It is observed that surface is damaged at jet entry; however it becomes smooth in middle. At jet exit, surface is rough along with striations. Wear tracks in the form of brittle fracture are observed in top region as shown in Figure 10. Also some damages are observed in the form of micro cracking and inter-granular fracture. This is due to low impact angle of AWJ. Also jet deflection results in micro cracking at this region. Smooth surface is observed in bottom region as shown in Figure 10. Some abrasive wear tracks are observed at this region due to stray abrasive particles. Minimum surface waviness is observed in this region due to steady AWJ. Long wear tracks are observed in bottom region of machined sample as shown in Figure 10. The damage of surface is observed along with micro cracks. Striations are observed at this region. This is due to the low kinetic energy of AWJ which tends to follow the path with least resistance.

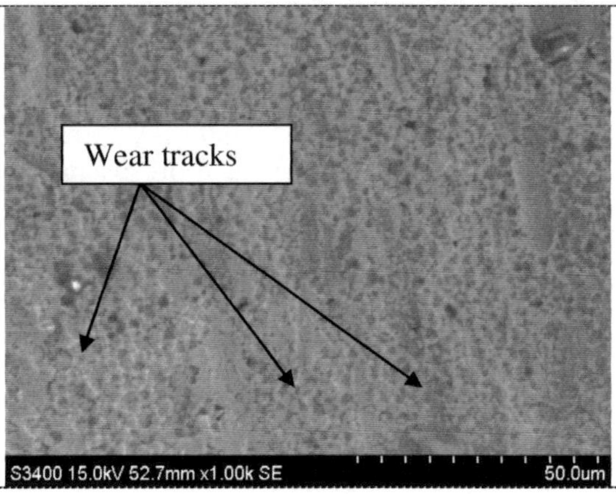

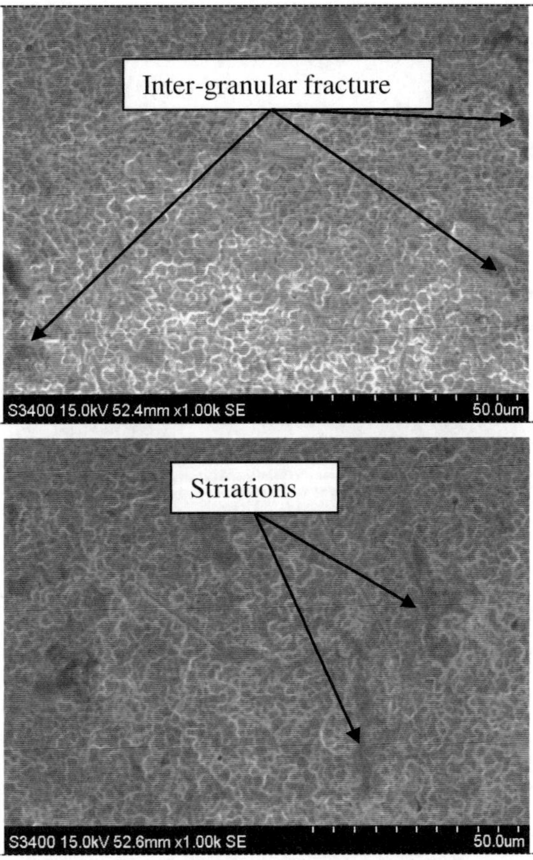

Figure 10. SEM image of top, middle and bottom region of machined surface

4. Multi-response optimization using desirability function

In the present study, a set of process parameters is optimized to minimize kerf taper and maximize depth of cut. The optimization is performed on the basis of desirability function. A dimensionless measure of response i.e. composite desirability function on the basis of desirability analysis technique is used for multiple responses. In this approach each response is converted into a desirability function (d_f) and it is bounded by $0 \leq d_f \leq 1$. The optimum values of stand-off distance, water pressure and traverse rate are obtained on the basis of statistical analysis in Design Expert v10 software. Optimized set of these process parameters gives a combination of levels which simultaneously fulfils optimization criteria for each response. Table 6 gives criteria for optimization.

Table 6. Criteria for Optimization

Factor	**Goal**	**Lower limit**	**Upper limit**	**Optimized value**
SOD	In range	2	4	2
WP	In range	140	160	160
TR	In range	310	330	310
Kerf taper	Minimize	1.1983	2.802	1.223
Depth of cut	Maximize	13.415	16.015	15.883

IOP Conf. Series: Journal of Physics: Conf. Series **870** (2017) 012019 doi:10.1088/1742-6596/870/1/012019

Bar graph for desirability of all factors and responses is shown Figure 11. This graph shows how well each factor satisfies the criteria and desirability of responses. Desirability value of all factors is 1. Desirability values of kerf taper, depth of cut and combined are 0.985, 0.949 and 0.967 respectively. Desirability variations with respect to WP and SOD are shown in Figure 12, when TR is at optimum level. Predicted value of desirability is 0.967 (approximately 1) which shows that process is most efficient at these operating conditions. Figures 13 and 14 show predicted values of kerf taper (1.223) and depth of cut (15.883) respectively at optimum levels. These figures are very useful to check how individual response behaves in the vicinity of optimum condition.

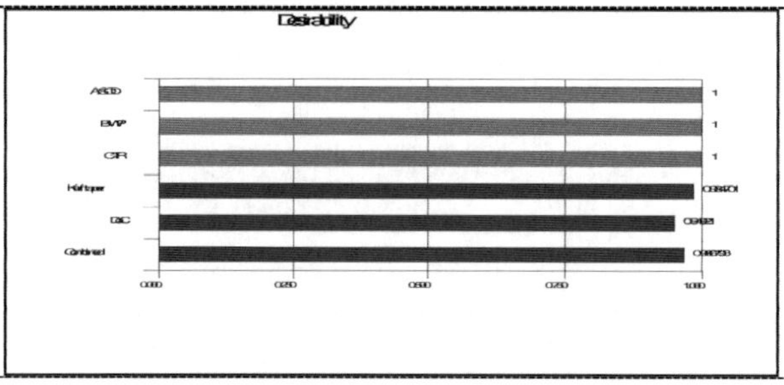

Figure 11. Bar graph of desirability

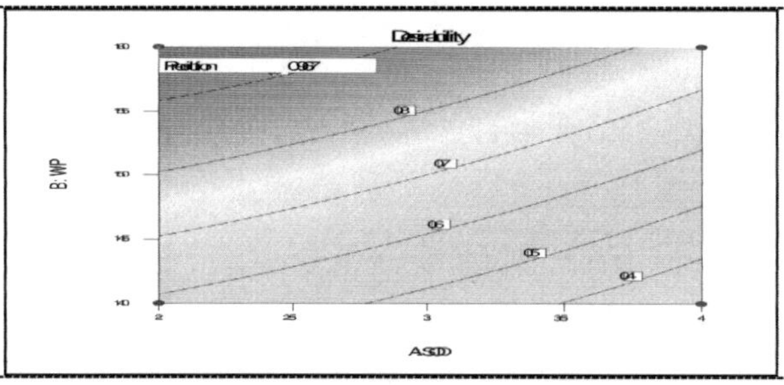

Figure 12. Desirability variations

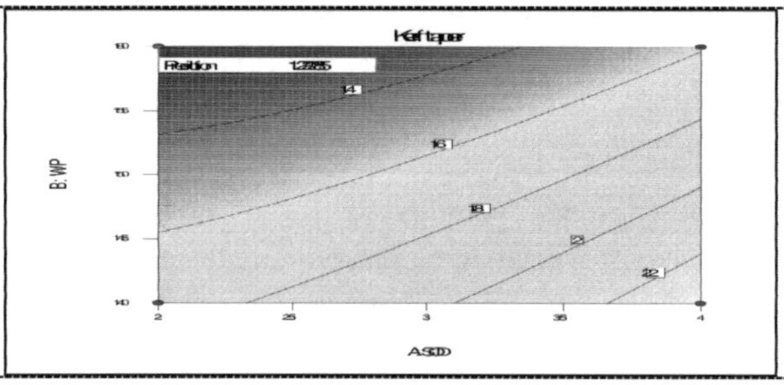

Figure 13. Predicted kerf taper at optimum levels

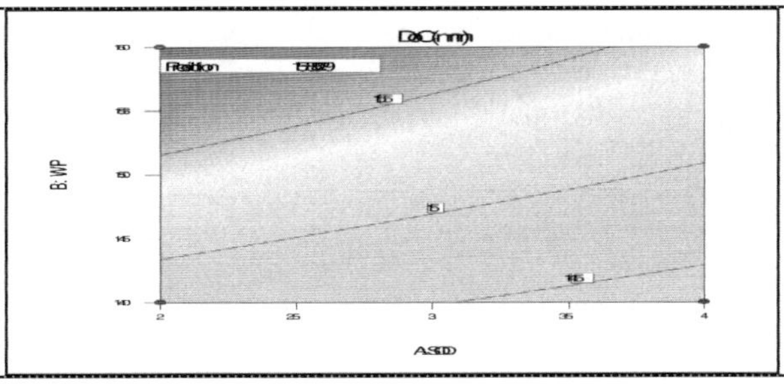

Figure 14. Predicted depth of cut at optimum levels

5. Conclusion

Plausible trends of kerf taper and depth of cut in AWJM of PZT ceramic have been analysed in the present work. Following are the findings of the present study-

(i) Traverse rate and water pressure are the most significant factors followed by stand-off distance to control kerf taper.

(ii) Kerf taper decreases with increase in water pressure and decrease in traverse rate and stand-off distance.

(iii) Water pressure is most significant parameter followed by traverse rate and stand-off distance to control depth of cut.

(iv) Depth of cut increases with increase in water pressure and decrease in stand-off distance and traverse rate.

Regression models are developed to predict kerf taper and depth of cut. It is found that model predictions are in good agreement with experimental results. The percentage error between predicted and experimental results is around 5. Further multi-response optimization of process parameters is performed to minimize kerf taper and maximize depth of cut.

References

[1] Barsoum M W, *Fundamentals of ceramics*, Institute of physics publishing (2003).

[2] CarterC B and NortonM G, *Ceramic Materials Science and Engineering,* Springer Science + Business Media New York, 4-31 (2013).

[3] Kiso H, TaguchiT, Fukuhara M and KimoraT, Bull. *Jpn. &w. Prec. Eng.,***21**, 142-144 (1987).

[4] Dhanawade A and Kumar S, *J. of Manufact. Eng*, **11,** 1-6 (2016).

[5] Wang J, *J. of Mate. Proc. Tech.*, **209**, 2314-2320 (2009).

[6] Hocheng H and Chang K R, *J. of Mate. Proc. Tech.,***40**, 287-304 (1994).

[7] Chen L, Siores E andWong W, *Int. J. of Mach. Tools and Manu.,***36**, 1201-1206 (1996).

[8] Momber A W, Eusch I and Kovacevic R, *J. of Mate. Sci.,***31**, 6485-6493 (1996).

[9] Gudimetla P, Wang J and Wong W, *J. of Mate. Proc. Tech.,***128**, 123–129 (2002).

[10] Wang J andGuo D M, *J. of Mate. Proc. Tech.,***133**, 371-377 (2003).

[11] Wang J and Liu H, *Proc. of the Inst. of Mech. Engrs., Part C,***220**, 715-725 (2006).

[12] Shanmugam D K and Masood S H, *Int. J. of Mach. Tools and Manu.,***48**, 1527-1534 (2009).

[13] Srinivasu D S and Axinte D A, *Proc. of the Inst. of Mech. Engrs., Part B,***225**, 317-335 (2010).

[14] Annoni M, Arleo F andMilano P, *Proc. of 21st Int. Conf. on Water Jetting: Looking to the Future, Learning from the Past*, Ottawa, Canada, 67-80 (2012).

[15] Huang C, Lv Z, Wang J, Zhu H, Yao P and Liu Z, Adv. Mate. Res., **797**, 15-20 (2013).

[16] Ghosh D, Das P K and Doloi B, *Proc. of 5th Int. and 26th Int. All India Manufacturing Technology, Design and Research Conf., Guwahati, India,* 422 (1-5) (2014).

[17] Dittrich M, Dix M, Kuhl M, Palumbo B and Tagliaferri F, *6th CIRP Int. Conf. on High Performance Cutting, Proc. CIRP,***14**, 442–447 (2014).

[18] Chithirai P S M, Sampath S S, Shetty S and ShivamurtyB, *Int. J. of Emerg. Tech. and Adv. Eng.,***5**, 402-407 (2015).

[19] Upadhyai R, Dhanawade A and Kumar S, *Proc. of 6th National conf. on Recent Adv. in Manu. Eng., Surat, India,* 255-260 (2016).
[20] Rouniyar A, Dhanawade A andKumar S, *Proc. of 6th National conf. on Recent Adv. in Manu. Eng., Surat, India,* 243-248 (2016).

IOP Conf. Series: Journal of Physics: Conf. Series **870** (2017) 012020 doi:10.1088/1742-6596/870/1/012020

Research on Attribute Reduction in Hoisting Motor State Recognition of Quayside Container Crane

F LI, G Tang and X Hu

Logistics Engineering College, Shanghai Maritime University, Shanghai 201306, China

gangtang@shmtu.edu.cn

Abstract. In view of too many attributes in hoisting motor state recognition of quayside container crane. Attribute reduction method based on discernibility matrix is introduced to attribute reduction of lifting motor state information table. A method of attribute reduction based on the combination of rough set and genetic algorithm is proposed to deal with the hoisting motor state decision table. Under the condition that the information system's decision-making ability is unchanged, the redundant attribute is deleted. Which reduces the complexity and computation of the recognition process of the hoisting motor. It is possible to realize the fast state recognition.

1. Introduction

As one of the important driving systems of quayside container crane, the hoisting motor directly influences the operation of quayside container crane. In order to keep abreast of the working state of the hoisting motor, the online condition monitoring system is often used to obtain the vibration signal of the hoisting motor, and then the state recognition. Mechanical equipment state recognition is essentially a pattern recognition or clustering problem, and the feature extraction of signal is one of the key elements. In order to comprehensively reflect the state of the motor, the multi-eigenvalues of the collected signals are usually extracted to form a high-dimensional characteristic parameter vector to evaluate the dynamic performance of the hoist motor comprehensively [1]. But too many characteristic parameters will take up too much storage space and affect the speed of calculation. At the same time, it will cause the problem of dimension disaster and characteristic redundancy [2]. Therefore, it is necessary to reduce the characteristic parameters of the hoisting motor to remove redundant information,

Rough set theory can remove the redundant information in the data, while maintaining the original classification ability [3]. In this paper, attribute reduction method based on rough set theory is used to attribute reduction of the high-dimensional eigenvector in the process of lifting recognition.

2. Basic definitions

Def 1. Let $\mathrm{IS} = (U, A, V, f)$ be an information system, where U is the set of objects, called the universe; A is the set of attributes, $\mathrm{A} = \mathrm{C} \cup \mathrm{D}$, C the condition attribute and D is the decision attribute; V is the set of values of attributes and f is an information function.

Def 2. For any subset, B belongs to A in the information system $\mathrm{IS} = (U, A, V, f)$, we can define the equivalence relation as follow.

IOP Conf. Series: Journal of Physics: Conf. Series **870** (2017) 012020 doi:10.1088/1742-6596/870/1/012020

$$IND_A(B) = \{(x,y) \in U^2 | \forall a \in B, a(x) = a(y)\}. \quad (1)$$

Where *a(x)* represents the value of object x on attribute a.

Def 3. Let R be a family of equivalence relations, $r \in R$, if $IND(R) = IND(R - \{r\})$, then r is unnecessary in R. Otherwise, r is necessary in R. If each $r \in R$ is necessary for R, then R is independent. Let Q ⊆ P, if Q is independent and IND (Q) = IND (P), then Q is called a reduction of P.

Def 4. The dependency of decision attribute on condition attribute is $\gamma_C(D)$.

$$\gamma_C(D) = |POS_C(D)|/|U| \quad (2)$$

Where $|U|$ is the cardinal number of universe, and $|POS_C(D)|$ is the positive domain of the decision attribute with respect to the condition attribute.

3. Attribute reduction method of hoisting motor information table

Discernibility matrix can be used to represent knowledge [4], this method has many advantages for the attribute kernel, attribute reduction and representation in the information table [5]. The attribute reduction algorithm based on discernible matrix and attribute selection is as follows.

Input: an information table of hoisting motor IS = (U, A, V, f)

Output: an attribute reduction of the information table RED(C)

Step 1. Calculated discernibility matrix of hoisting motor information table;

Step 2. The attribute core, CORE(C), of the information system is obtained according to the discernibility matrix, let B=CORE(C);

Step 3. For any $c_{ij} \epsilon M(i,j = 1,2,\cdots.n)$, if $c_{ij} \cap B \neq \emptyset$, let $c_{ij} = 0$;

Step 4. If any $c_{ij} = 0(i,j = 1,2,\cdots.n)$, skip to *step 6*, otherwise execute *step 5*;

Step 5. The number of occurrences of each attribute in the current discernible matrix is counted, and the most frequent elements are selected, B ⇐ B ∪ $\{\alpha\}$, then go to the *step 3*;

Step 6. Output the result B ∈ RED(C).

4. Attribute reduction method of decision table

Solving all attribute reductions or minimal reductions of a decision table has proved to be an exponentially difficult problem [6]. Therefore, this paper uses the combination of rough set theory and genetic algorithm to reduce the redundancy condition attributes [7], and find the global optimal reduction of the decision table of the hoisting motor through the global optimization probabilistic search algorithm of genetic algorithm. Specific algorithm steps are as follows:

Input: a decision table of hoisting motor

Output: an optimal attribute reduction of the decision table

Step 1. Calculate the support degree of D with respect to C.

Step 2. Set the fitness function

$$\mathrm{F(r)} = (l - l_r)/l + \gamma_C(D) \quad (3)$$

Step 3. Calculate fitness of each individual in the initial population

Step 4. Generate new population according to crossover probability and mutation probability.

Step 5. Calculate fitness of each individual in the new population

Step 6. Copy the optimal individual into next generation according to the optimal preservation *strategy*, and save the optimal individuals.

Step 7. If the best individual fitness value of continuous n generation is no longer increased output the optimal individual, otherwise turn back *Step 4.*

5. Attribute reduction of hoisting motor information table

The state identification of the hoist motor is generally an unsupervised clustering process, that is, the state of the hoist motor is not known in advance [8]. Therefore, there is no decision attribute in the information table. Using the acceleration sensor mounted on the output shaft of the hoisting motor to collect the vibration signal, and then extract the various characteristic parameters of the vibration signal to form the condition attribute set of the lifting motor information table. But not all of the condition attributes are necessary for state recognition. In order to achieve fast clustering and reduce computational cost, it is necessary to attribute reduction and delete redundant attribute of hoisting motor information table.

Taking the online condition monitoring data of the hoisting motor of 4 weeks as the study object, we take the day unit as the research object, $U = [1, \ 2, \ \cdots, \ 28]'$. The daily mean value of the vibration signal, the square root amplitude, the skewness index, the waveform index, the peak value index, the pulse index and the margin index are calculated respectively [9]. A total of 8 special diagnosis parameters are taken as the condition attributes of the hoisting motor state information table. Respectively, with $c_1, c_2, c_3, c_4, c_5, c_6, c_7, c_8$ said. The obtained eigenvalues are discretized (using the method of equal frequency discretization in this paper) to obtain the lifting motor state information table as shown in Table 1.

Table 1. Hoist motor status information table of hoisting motor

U	c_1	c_2	c_3	c_4	c_5	c_6	c_7	c_8
1	3	3	1	1	1	1	1	1
2	3	3	1	1	2	1	1	2
3	1	1	3	3	2	3	3	2
4	2	3	1	1	1	1	1	1
5	2	2	2	2	2	1	2	2
6	1	1	3	3	3	3	3	3
7	3	3	1	2	1	1	1	1
8	3	2	2	2	2	2	2	2
9	2	2	2	2	3	3	3	3
				...				

The attribute reduction method based on discernibility matrix and attribute selection is used to attribute reduction of the hoisting motor information table. The results show that the $\mathrm{RED}(C) = \{c_1, c_2, c_3, c_4, c_6\}$ is the effective value, square root amplitude, skewness index, kurtosis index and peak value. After deleting the redundant attribute in the information table, the reduced feature vector has only 5 dimensions. Which is reduced by 38% compared with that before reduction, which greatly reduces the computational cost and provides a guarantee for realizing fast state identification and visualization.

To a certain extent, the motor vibration signal reflects the running status of the motor. For example, when the peak value of the vibration signal is too high, it indicates that the motor is not running smoothly or is overloaded [10]. The hoisting motor vibration signal for low-pass filter processing to remove noise and other interference signals. And then takes the peak value of the vibration signal after filtering as the decision attribute D of the state of the hoisting motor. The decision table of the hoisting motor is discretized (the method of equal frequency discretization is adopted here), and the decision table of the hoisting motor status is shown in Table 2. D= {1,2,3}, respectively, according to the different motor vibration signal peak divided into three states of the lift motor.

Table 2. Decision table of hoisting motor

U	c_1	c_2	c_3	c_4	c_5	c_6	c_7	c_8	D
1	3	3	1	1	1	1	1	1	3
2	3	3	1	1	2	1	1	2	3
3	1	1	3	3	2	3	3	2	1
4	2	3	1	1	1	1	1	1	3
5	2	2	2	2	2	1	2	2	3
6	1	1	3	3	3	3	3	3	2
7	3	3	1	2	1	1	1	1	3
8	3	2	2	2	2	2	2	2	3
9	2	2	2	2	3	3	3	3	2
				…					

In order to find the decision rules between the condition attributes and the decision attributes of the hoist motor, the rough set theory and the genetic algorithm are used to reduce the attributes of the decision table. The dependency of decision attribute on condition attribute is added to the fitness function to ensure that the condition attribute has sufficient support for the decision attribute after reduction. The parameters of the genetic operator are set to be Pc = 0.7, Pm = 0.01 and m = 15. The optimal individual after reduction is {1 0 0 0 1 0 1 0}, i.e. the condition attribute {c_1 c_4 c_7}, And the results after reduction are shown in Table 3.

Table 3. Reduction result

c_1	c_4	c_7	D
3	1	1	3
3	2	1	3
1	2	3	1
2	1	1	3
1	3	3	2
3	2	2	3
2	3	3	2
1	1	2	1
2	2	2	2
1	1	1	1
3	1	2	3
2	3	2	3

From Table 3 can be seen to determine the state of the hoisting machine is mainly RMS, kurtosis index and pulse index. The effective value reflects the fluctuation of the signal relative to the zero value, and the effective value corresponds to the vibration energy. According to the generalization standard, it can be found that: when the effective value is relatively large, the hoisting motor vibration is relatively large; when the effective value is relatively small, the hoisting motor vibration is relatively small; RMS and hoisting motor has a similar related relationship. The kurtosis index is a dimensionless parameter, which is independent of the structure of the lift motor and is particularly sensitive to the shock signal, and the pulse index is also a dimensionless index, which is characterized by being sufficiently sensitive to the running state of the hoist motor. When the motor running state changes caused by the vibration signal changes, the pulse index has changed significantly. And there is a stable correspondence between the pulse index and the running state of the hoisting motor, and it is insensitive to the factors other than the running state of the hoisting motor, such as the magnitude of the load.

6. Summary

In the process of state recognition of the hoisting motor of the quayside container crane, the characteristics of the feature parameter are too many and the redundant attribute affects the efficiency of state recognition. In this paper, an attribute reduction method based on rough set theory is used to deal with hoisting machine state information table. Under the premise of guaranteeing the state recognition rate and the recognition precision, the lifting machine state information table is reduced in dimension. Through attribute reduction, redundant attributes such as waveform index and margin index in hoist motor state information table are deleted. So as to achieve the purpose of reducing the complexity and the calculation amount of the lifting motor state identification process. The algorithm of attribute reduction based on the combination of genetic algorithm and rough set theory is also introduced in detail. Based on the algorithm, the attribute reduction of the decision table of the hoist motor is made, and the decision relation between the effective value, the kurtosis index and the pulse index and the hoist motor state is found.

Acknowledgements

This work was supported by the National Natural Science Foundation of China (No. 31300783), China Postdoctoral Science Foundation (No. 2014M561458), Doctoral Fund of the Ministry of Education Jointly Funded Project (No. 20123121120004), the Shanghai Maritime University Research Project (No. 20130474), the Shanghai Top Academic Discipline Project- management science & engineering, and the high-tech research and development program of China (No. 2013A2041106).

References

[1] Hu X, Wang Z X and Cheng Z N 2007 Research on features reduction in mechanical condition recognition about motors based on rough set *Journal of Vibration and Shock.* **26** 32
[2] Milligan G W 1980 An examination of the effect of six types of error perturbation on fifteen clustering *J. Psychometrika.* **45** 325
[3] Pawlak Z 1982 Rough sets *International Journal of Parallel Programming.* **38** 341
[4] Skowron A 2010 Extracting laws from decision tables: A rough set approach *J. Computational Intelligence.* **11** 371
[5] Miao D Q and Li D G 2008 *Rough sets theory algorithms and applications* (Beijing: Tsinghua University Press) p 28
[6] Wong S K M 1985 Optimal decision rules in decision table *J. Bulletin of Polish Academy of Sciences.* **33** 693
[7] Liu B and Chen D J 2010 Analysis of the traffic accidents based on rough set theory and genetic algorithm *J. Journal of Lanzhou Jiaotong University.* **29** 69
[8] Chakhar S, Ishizaka A, Labib A and Saad I 2016 Dominance-based rough set approach for group decisions *European Journal of Operational Research* **251** 206
[9] Ishizaka A and Nemery P 2013 *Multi-Criteria Decision Analysis: Methods and Software.*
[10] Figueira J, Greco S and Ehrogott M 2005 Multiple criteria decision analysis: state of the art surveys *International* **142** 192

Finite Element Analysis of Doorframe Structure of Single Oblique Pole Type in Container Crane

X. F. Cheng[1a], F. Q. Wu[2b], G. Tang[1c*], X. Hu[1d]

[1]Logistics Engineering College, Shanghai Maritime University, 201306, Shanghai China
[2]Shanghai Institute of Special Equipment Inspection & Technical Research, 200062, Shanghai China

[a]dafeng1121@126.com ,[b]81414458@qq.com ,[c]gangtang@shmtu.edu.cn,[d]huxiong@shmtu.edu.cn

Abstract. Compared with the composite type, the single oblique pole type has more advantages, such as simple structure, thrift steel and high safe overhead clearance. The finite element model of the single oblique pole type is established in nodes by ANSYS, and more details are considered when the model is simplified, such as the section of Girder and Boom, torque in Girder and Boom occurred by Machinery house and Trolley, density according to the way of simplification etc. The stress and deformation of ten observation points are compared and analyzed, when the trolley is in nine dangerous positions. Based on the result of analysis, six dangerous points are selected to provide reference for the detection and evaluation of container crane.

1. Introduction

Container crane is the most important equipment for container special port [1,2]. Super-speed and large-scale are the future direction of container crane. The rated load under the spreader has increased gradually from 30.5t to 61.5t, and the maximum has reached 65t [3]. With outreach and lifting height increasing, lifting speed is enhanced from 50 / 120m / min to 90 / 200m / min and so on. These constraints put forward higher requirements for the steel structure of container crane. Finding the weaknesses of the structure is significant for container crane' safety. Optimization and detection of these places are also equally important [4]. This paper uses ANSYS software to analyze the common single oblique pole type. The stress and deformation of observation points are analyzed and compared when Trolley is at different dangerous positions. Six dangerous points are chosen by the results of the analysis finally.

2. The finite element model of the steel structure is established

2.1. Simplification and hypothesis of model

Container crane is complex generally [5]. If all the details are taken into account, it is bound to increase the difficulty of modeling and the result is not necessarily well. Therefore, the container crane should be simplified reasonably and ideally. A finite element model is established, which is both beneficial for the calculation and can be accordance with the actual working condition [6].

IOP Conf. Series: Journal of Physics: Conf. Series **870** (2017) 012021 doi:10.1088/1742-6596/870/1/012021

(1) The whole structure includes the Leg Landside(LS.) and Leg Waterside(WS.), Frame, Boom, Girder etc. As these parts are welded by the steel plate, so the beam element is used to model [7];

(2) As the Machinery house is fixed on the Girder by several legs, the Machinery house is simplified to the mass points on the Boom. Torque occurred by the Machinery house is also applied on the Boom. The trolley is simplified into four mass points. Torque occurred by the Trolley is considered at the same time;

(4) The Gantry Operating agencies, main Hoist agencies etc, are simplified to mass points imposed in corresponding locations;

(5) Ladders, Rails, Ribs, etc. have no effect on the whole structure, so the density of corresponding components are adjusted. These accessories are not constructed in the model;

(6) As the cross-section of the Boom and Girder are complex, a custom cross-section is chosen. The track is a part of Girder and Boom;

(7) The Gantry is simplified into four legs;

(8) The ear plates are simplified as beam connected Boom, Girder, Forestay and Backstay;

(9) When the Boom is level, Forestay is deemed as a tension and compression to withstand the two-pole, therefore LINK180 is used to model.

2.2. *The element type of the model* [8]

(1) BEAM188: Frame, Boom, Girder and other connection systems. The Frame and the connection systems are used rectangular sections; the Boom and Girder use custom sections;

(2) PIPE288: Backstay, the support system between the Frame, Boom, Girder and Pylon;

(3) MASS21: The Machinery house and some other additional accessories;

(4) LINK180: Forestay.

2.3. *Material properties of the model* [9,10]

(1) Mechanical properties： Steel structure use generally A709-50-1 steel, the mechanical properties of elastic modulus E = 2.1GPa, Poisson ratio μ = 0.3;

(2) Material density: Frame, Girder and Boom have a large number of internal ribs, the external escalator, Forestay and Pylon have pins, shafts, baffles and other parts etc. These have been simplified as discussed upon. Densities of these accessories should be adjusted to balance the model weight and the actual weight.

2.4. *Model Constraint and Condition*

Constraint: In the ANSYS model, the Gantry are simplified into four support legs, so that the four bottom nodes' the x, y, z direction of displacement are constrained [6]. According to another papers, 9 dangerous positions (as shown in Table 1, Figure1) be selected where Trolley are on the Boom and Girder to analyze and calculate.

Table 1. Nine dangerous positions

NO.	Dangerous positions
P1	In the Backreach
P2	In the middle of the Tailstay and the Backreach
P3	In the Portal upper beam LS.
P4	In the middle of the Portal upper beam WS. and Portal upper beam LS.
P5	In the Portal upper beam WS.
P6	In the middle of Forestay's hinge and Boom's hinge
P7	In Forestay's hing
P8	In the middle of Forestay's hings

IOP Conf. Series: Journal of Physics: Conf. Series **870** (2017) 012021 doi:10.1088/1742-6596/870/1/012021

P9 Max-Outreach

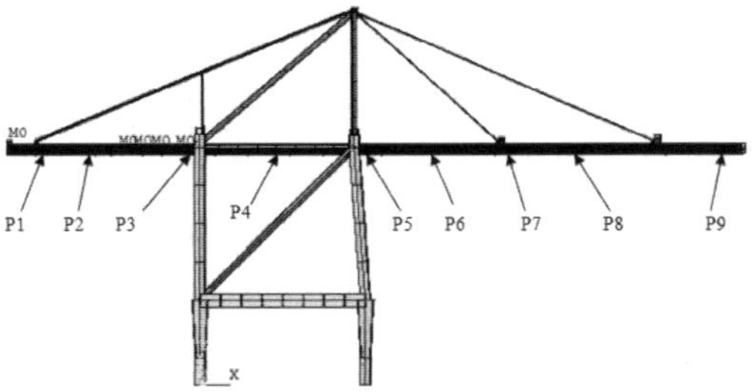

Figure1.Dangerous positions

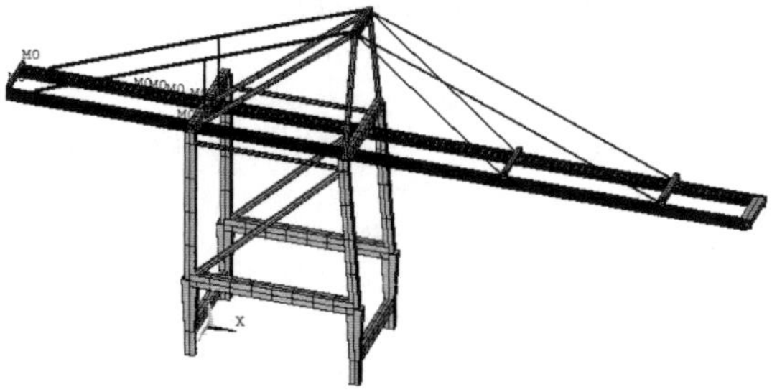

Figure2. Finite model

2.5. *Model establishment*

Traditional modeling is generally used bottom-up modeling or other software to draw a solid model and then import the ANSYS software. The finite element software analyzes nodes and elements ultimately, so at the beginning of modeling, nodes and elements are chose to establish model, without other objects[7]. The procedure as follows: select the element - input material properties - create cross-section of beam, pipe and custom sections, link sectional area and mass elements - create node and element - couple the units - adjust the density - load calculation - impose constraints and loads[11]. The established model is shown in Figure2. It should be defined that length unit mm, mass unit t, force unit N, density unit t / mm ^ 3, the elastic modulus unit MPa, the acceleration unit mm / sec ^ 2.

3. Load calculation

3.1 Main working load.

According to the literature, the loads acting on the crane are divided into four categories: normal load, incidental load, special load and other loads in the working state and the non-operating state[12]. It is assumed that all the moving parts are at their most unfavorable positions, the main working load of the container crane is：

(1) Container crane overall weight PC;

(2) Container rated weight PQ,Trolley weight P1， Spreader weight P2;

(3) The hoisting load caused by the working load: the fluctuating dynamic coefficient ψ resulting from the lifting of the working load is multiplied by the load due to the working load, where $\psi = 1 + \xi VL$[13] , ξ=0.3.The model comes from Zhen hua Port Machinery Company (ZPMC) . Lifting speed(VL) is 1.12m / sec, this formula requires 1m / sec as maximum lifting speed. When the speed is greater than 1m / sec, the dynamic coefficient ψ is no longer increased;

(4) The level of inertia induced by the movement of Trolley: The formula used for the simplified calculation is F = 2ma, where 2 is the load increase coefficient, m is the sum of the Trolley weight, the Spreader weight and the Container weight and a is the Trolley starting acceleration or braking deceleration [13].

3.2 Load combination.

According to the working characteristics of the container crane, in this calculation, the metal structure of the container crane belongs to the statics characteristic, the load combination is:

(1) The vertical load (y direction): $DP1 = rc \times [\psi \times (P2 + PQ) + P1]$, where rc is the increase factor, according to the European crane design specification value is 1.2;

(2) Horizontal load (x direction): $F=2ma$.

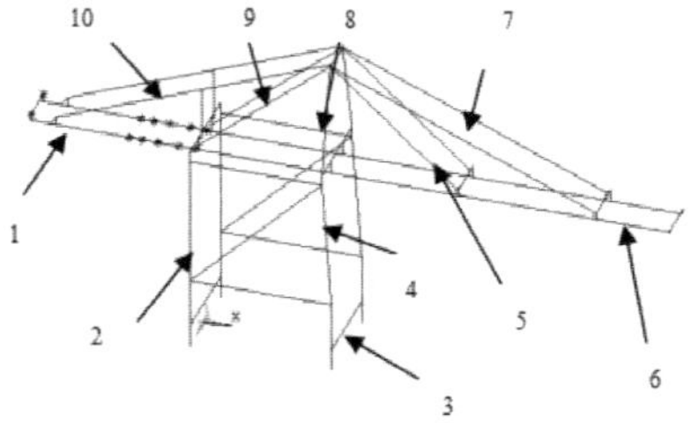

Figure3. Main parts

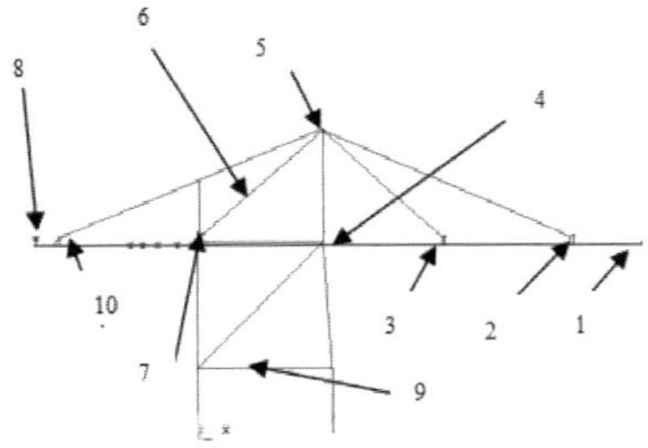

Figure4. Deformation observation points

4. Finite element analysis calculation

4.1 Static analysis of steel structure of container crane

(1) It is necessary to analyze the stress of each part of the container crane in order to understand the stress of the various parts of the container crane[14]. In this paper, 10 important parts of the container crane are selected and numbered, showed in Pigure3.The stress analysis results of various components is shown in Table 2. In Figure5, it can be drawn that the Girder, Boom, Pylon, Leg WS. and the inside Backstay are the most sensitive stress places;

(2) Vertical deformation analysis: According to the literature, 10 deformation observation points are selected, and the distribution and number are shown in Figure4. The vertical deformation of the observation points are shown in Table 3, where the negative means downward. In Figure6 ,it can be drawn that the Girder and Boom end point, outside Forestay hinge point, inside Forestay hinge point, Backstay hinge point and the midpoint of the top beam, the six points are the most sensitive deformation points.

Table.2 Stress of main parts in different conditions of container crane steel structure

No.	Part Name	Stresses of various parts of the metal structure of the quayside bridge under load (MPa)								
		P1	P 2	P 3	P4	P 5	P6	P 7	P8	P 9
1	Girder	204.7	209.7	142.7	152.9	128.6	220.4	124.2	124.7	122.7
2	Leg LS.	102.9	95.1	64.8	47.1	40.8	29.8	19.9	15.4	24.8
3	Sill beam WS.	5.7	5.7	5.6	5.1	4.2	5.2	5.4	5.9	5.8
4	Leg WS.	50.9	53.8	59.3	69.2	91.6	89.7	90.8	91.4	105.7
5	Boom	46.7	46.7	46.9	47.1	46.7	222	53	112.1	150.4
6	Inside Forestay	52.9	53	53.3	53.7	52.7	51.8	55	77	135
7	Outside Forestay	57.1	56.8	56	55	57.5	92	123	118	51
8	Pylon WS.	81.5	77.7	74.6	75.3	77.2	87.1	101.7	114.7	129.3
9	Inside Backstay	69.5	73	70.1	63.1	69.9	79.9	88.1	98.4	114.2
10	Outside Backstay	115	92.9	58.7	52.6	57.6	59.9	59.1	59.2	59.8

Table.3 Deformation of transform observation points in different conditions

No.	Observation point	Deformation values of the deformation observation points (mm)								
		P1	P2	P3	P4	P5	P6	P7	P8	P9
1	Boom end	-60	-75.9	-107.4	-123.9	-129.7	-149.7	-174.7	-213.5	-407
2	Outside Forestay hinge point	-85.5	-85.5	-85.5	-85.5	-85.5	-85.5	-85.5	-85.5	-85.5
3	Inside Forestay hinge point	-17.2	-23.6	-35.8	-42.1	-46.3	-68.8	-87.8	-94.7	-95.3
4	Below Portal WS.	-10.3	-10.9	-13.2	-17.4	-14.2	-16.4	-17.8	-16.9	-19

5	Midpoint of the top beam	-12.2	-13.1	-14.6	-15.6	-16.4	-18.8	-21.2	-23.5	-26.3
6	Inside Forestay	-10.2	-10.7	-14.3	-16.9	-14.2	-13.7	-15	-16.1	-17
7	Portal LS.	-25.1	-23.4	-19.8	-16.4	-13	-9.2	-6.6	-3.0	2.7
8	Girder end	-151.3	-116.8	-62.7	-44.5	-45	-29.3	-11.6	9.2	38.9
9	Horizontal beam	-4.9	-4.8	-4.6	-4.4	-4.5	-4.5	-4.5	-4.4	-4.3
10	Outside Backstay hinge point	-161.7	-130.6	-71.1	-52	-54.5	-41.8	-26.1	-8	-17.8

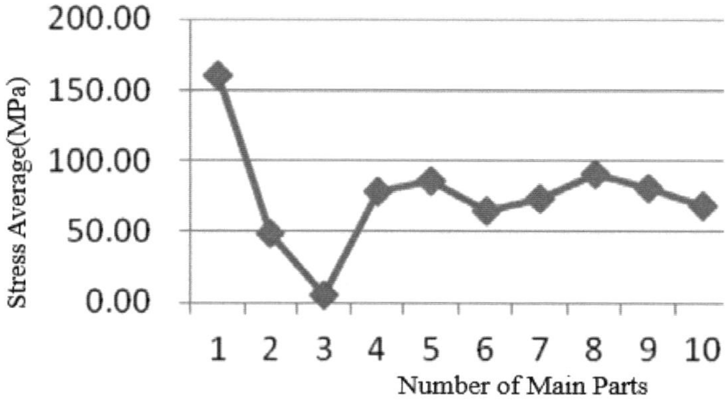

Figure5. The average stress of measuring parts

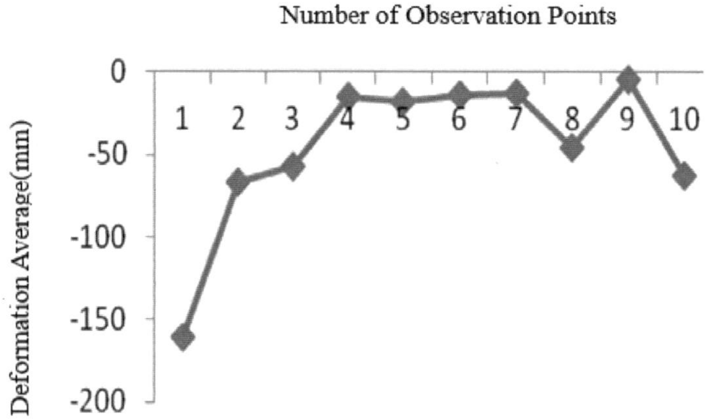

Figure6. The average deformation of observation points

5. Summary

In this paper, the finite element model of single oblique pole type in container crane is established by the ANSYS software, and the stress and deformation are analyzed for the important parts and observation points under different working conditions. We can conclude that:

(1) The Girder, Boom, Pylon, Leg WS. and the inside Backstay are the most sensitive stress places;

(2) The Girder and Boom end point, outside Forestay hinge point, inside Forestay hinge point, Backstay hinge point and the midpoint of the top beam, the six points are the most sensitive deformation points;

(3) In the process of detecting and diagnosing the container crane, we can simply pay attention to these six deformation observation points and these five parts.

Acknowledgements

This work was supported by the National Natural Science Foundation of China (No. 31300783), China Postdoctoral Science Foundation (No.2014M561458), Doctoral Fund of the Ministry of Education

Jointly Funded Project (No.20123121120004), the Shanghai Maritime University Research Project (No.20130474), the Shanghai Top Academic Discipline Project- management science & engineering, and the high-tech research and development program of China (No. 2013A2041106)

References

[1] Dong Dashan, Mei Xiao, Zheng Su. *Port Crane*.(Shanghai: Shanghai Jiao tong University Press), 2014pp19-35.
[2] Yang Yuhua, Zhang Qing, Nie Feilong. *Development Trend of Container Automation Terminal,*(Shanghai: Chinese Journal of Construction Machinery),2015p571.
[3] Zhu Yongjun. *Quayside Container Crane Structure's Parameterize Analysis and the Development of the Software System*, (Wuhan: Wuhan University of Technology),2014pp1-6.
[4] Du Chao, Wang Jiaoe, Mo Huihui. *Research on the Spatial Pattern and Complexity of Chinese Container Shipping Network*,(Wuhan: Resources and Environment in the Yangtze Basin),2016.
[5] Huang Xiaojie. *Research on Stress and Fatigue Life Analyze and Online Visualization Technology for Container Crane Component*,(Shanghai: Shang Hai Jiao TongUniversity),2014
[6] Deng Guoping, Wu Zhijun, Qiu Huiqing.*Finite element analysis and comparison of the gantrycrane mechanism for the quayside container cranes*,(Shanghai:Chinese Journal of Construction Machinery), 2011pp43-47.
[7] Cheng Feng. *Based on ANSYS shore crane parameterized simulation and fatigue analysis*,(Wuhan: Wuhan University of Technology), 2007pp1-48.
[8] Li Guofang. *Optimization Design of Main Girder of Container Crane*,(Beijing: China Science and Technology Information),2014pp152-153.
[9] Shang Yuejin.*Finite theory and ANSYS application guide*,(Beijing: Tsinghua University Press), 2005.
[10] Dietrich Braess. *Finite Elements*.Beijing: World Book Inc,2012.
[11] Li Jing. *Research on Load Spectrum and Fatigue Life of the Container Crane*,(Wuhan: Wuhan University of Technology),2014.
[12] National *Standard of the People 's Republic of China GB / T 3811-2008 Crane Design Code.*
[13] Federation *European Manufacture， FEM 1.001： European Crane Design Specification.*
[14] Ta Na, Rao Zhusi, Li Yiming. *Finite Element Modeling and Dynamic Response Analysis of Container Crane*,(Shanghai: Noise and Vibration Control),2008.

IOP Conf. Series: Journal of Physics: Conf. Series **870** (2017) 012022 doi:10.1088/1742-6596/870/1/012022

State Recognition and Visualization of Hoisting Motor of Quayside Container Crane Based on SOFM

Z.Q. Yang[1a], P. He[2b], G. Tang[1c*], X. Hu[1d]

[1]Logistics Engineering College, Shanghai Maritime University ,1550 Harbor Road, Pudong New Area, 201306, Shanghai, China
[2]Shanghai East Container Terminal, 200137, Shanghai,China

[a]youngzhiqi111@163.com,[b] hep@sect.com.cn, [c]gangtang@shmtu.edu.cn,[d]huxiong@shmtu.edu.cn

Abstract. The neural network structure and algorithm of self-organizing feature map (SOFM) are researched and analysed. The method is applied to state recognition and visualization of the quayside container crane hoisting motor. By using SOFM, the clustering and visualization of attribute reduction of data are carried out, and three kinds motor states are obtained with Root Mean Square(RMS), Impulse Index and Margin Index, and the simulation visualization interface is realized by MATLAB. Through the processing of the sample data, it can realize the accurate identification of the motor state, thus provide better monitoring of the quayside container crane hoisting motor and a new way for the mechanical state recognition.

1. Introduction

Before mining monitoring information of the quayside container crane hoisting motor, the priori knowledge of the data often fails to be gained. A state-recognition which is actually clustering analysis problem is an effective method, and unsupervised learning clustering algorithm is used to solve the problem.

The state recognition of multi-feature parameters of quayside container crane hoisting motor can obtain the sample data through multiple sensors to form a multi-dimensional space, and then the data is reduced and recognized to the low-dimensional space. In this mapping process, the low-dimensional space mapping should preserve and highlight the main properties in the original high-dimensional data, and thus to improve the accuracy of motor state recognition.

2. High-dimensional data visualization and its methods

High-dimensional data usually refers to the variable form that is $F=f\left(x_1,x_2,L\ x_n\right)$. For example, there are N monitoring points in the quayside container crane hoisting motor, and each monitoring point has 9 eigenvalues to form the N multiplied 9 dimensional data. Visualization of high dimensional data is mostly realized by dimensionality reduction, Linear or nonlinear methods can be used to transform high-dimensional space into two-dimensional or three-dimensional space that human can perceive[1].

The SOFM algorithm is an unsupervised clustering method. The network can repeat the learning of the input pattern so that the probability distribution of the connection weight vector is consistent with the input pattern. It has two layers of network[2-6], one input layer, the other layer includes the competition layer and the output layer. and there are two typical structures including one-dimensional linear array and two-dimensional linear array.

3. Clustering Analysis of Mechanical State of the quayside container crane hoisting motor

3.1 Data reduction.

In the monitoring of the hoisting motor, several acceleration sensors are placed, and nine characteristic values are extracted from the collected signals, such as Square root amplitude of the vibration velocity, the effective value, the skewness index, the kurtosis index, the peak value index, the pulse index, the margin index and the variance index. Taking 28 days data of the quayside container crane hoisting motor, the relevant characteristic data are obtained, and 4000 samples of the data are collected every day to be analyzed and visualized. The following results are obtained by attribute approximation and normalization.

Table 1. Reduction results of the eigenvalues

Time	Margin index	RMS	Pulse index
1	-0.75619	0.77337	-0.97058
2	-0.52976	0.78704	-0.93672
3	-0.54628	-1.00000	-0.47828
4	-0.70062	0.47155	-0.95843
5	-0.42726	0.46136	-0.91232
6	-0.27790	-0.04987	-0.68793
7	-0.62404	0.73640	-0.93803
8	-0.57819	0.54370	-0.93822
9	-0.41272	0.50735	-0.88982
10	-0.23475	0.27405	-0.77565
11	-0.89796	-0.97168	-0.85472
12	0.15387	0.03130	-0.69016
13	-0.47880	1.00000	-0.94729
14	-0.52498	0.26793	-0.86347
15	-0.65586	0.56287	-0.91602
16	-0.66216	0.93173	-0.98850
17	1.00000	-0.83678	1.00000
18	0.35139	-0.81156	0.19050
19	-1.00000	-0.96401	-1.0000
20	-0.10624	0.15986	-0.78445
21	-0.61372	0.50796	-0.91464
22	-0.50187	0.46266	-0.89533
23	-0.24473	0.49439	-0.83101
24	-0.49887	0.54367	-0.90312
25	-0.01591	-0.13438	-0.58409
26	-0.15392	0.30469	-0.79925
27	-0.45777	0.33085	-0.87283
28	-0.40532	-0.08769	-0.70611

The SOFM algorithm is realized by MATLAB. The topological structure of the network is constructed and selected. Six topological structures are constructed. The order of neurons is 1×3、2×3、3×3、4×5、4×7. $\eta(t) = 0.1$ Each network structure achieves a certain number of training, so that the network structure tends to be stable after training[7-9].

From the training topology of network and output results, some neurons can't be fully utilized[10], when there are too many neurons. In order to achieve more accurate network mapping, the appropriate neuron structure needs to be selected. The data can be divided into 3 to 9 kinds. In order to express it more accurately, the network topology arranged 3×3 is selected for data analysis.

IOP Conf. Series: Journal of Physics: Conf. Series **870** (2017) 012022 doi:10.1088/1742-6596/870/1/012022

3.2 Data clustering results.

After normalization of the reduced data, the SOFM algorithm is used to deal it with [11], $\eta(t)=0.1$ and the number of training net.trainParam.epochs = [100 500 1000 2000 3000 4000] is tested in Fig.1.

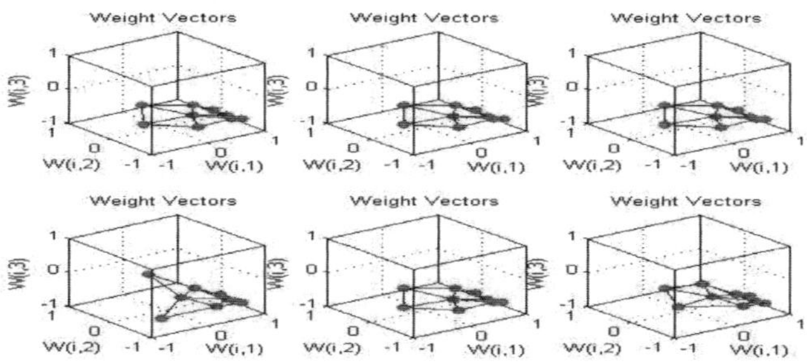

Figure 1. Network topology after different training epochs

To avoid over-fitting, as can be seen from the figure above, the network topology of 3000 training is selected, and then deal with the data. $\eta(t)=0.1$ Figure 4 and Table 2 are obtained.

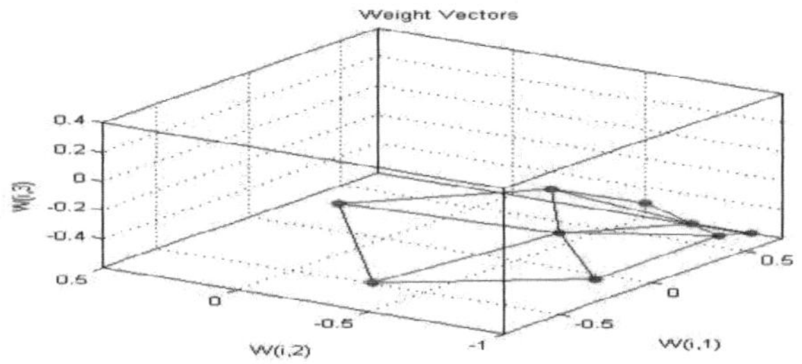

Figure 2. Topology of training network

Table 2. Clustering situation

Time	A	B	C	D	E	F	G	H	I
1	1	0	0	0	0	0	0	0	0
2	1	0	0	0	0	0	0	0	0
3	0	0	0	0	0	1	0	0	0
4	0	1	0	0	0	0	0	0	0
5	0	0	0	1	0	0	0	0	0
6	0	0	0	0	1	0	0	0	0
7	1	0	0	0	0	0	0	0	0
8	1	0	0	0	0	0	0	0	0
9	0	0	0	1	0	0	0	0	0
10	0	0	0	0	0	0	1	0	0
11	0	0	1	0	0	0	0	0	0
12	0	0	0	0	0	0	0	1	0
13	1	0	0	0	0	0	0	0	0
14	0	1	0	0	0	0	0	0	0
15	1	0	0	0	0	0	0	0	0
16	1	0	0	0	0	0	0	0	0
17	0	0	0	0	0	0	0	0	1
18	0	0	0	0	0	0	0	0	1
19	0	0	1	0	0	0	0	0	0
20	0	0	0	0	0	0	1	0	0

21	1	0	0	0	0	0	0	0	0
22	0	0	0	1	0	0	0	0	0
23	0	0	0	1	0	0	0	0	0
24	1	0	0	0	0	0	0	0	0
25	0	0	0	0	0	0	0	1	0
26	0	0	0	0	0	0	1	0	0
27	0	0	0	1	0	0	0	0	0
28	0	0	0	0	1	0	0	0	0
Frequency	9	2	2	5	2	1	3	2	2

The clusters can be analyzed in the graph and the table, mainly focus on A, D, and G classes, the range of their characteristic vector margin index, RMS and pulse index as follows:

Table 3. Clustering state index

Class	Margin index	RMS	Pulse index
A	1.5795±0.1	2.1278±0.1	8.6650±1
D	1.7915±0.1	1.8328±0.2	10.3957±1
G	2.0625±0.2	1.5986±0.1	13.1490±2

It can be seen from the table that a group of eigenvectors of each state has its own characteristics. Within a month, the quayside container crane hoisting motor is in normal range and does not fail, so this difference of the clustering index value is not big. So, the SOFM algorithm can achieve very accurate state recognition and visualization, which can monitor the mechanical state better. Then it should be paid special attention to the state of this machine, and timely inspect and repair when the eigenvector index difference is too big in the future.

4. Conclusion

The processed data is based on the day as a unit, and then carried out the motor status identification and visualization to achieve a very accurate clustering and visualization by using self-organizing feature map (SOFM) algorithm. The processed data is also based on the hour as a unit to achieve a more accurate analysis. Through the data acquisition and processing of the hoisting motor, it is possible to accurately identify the working state of the motor, thus provide better monitoring of the hoisting bridge motor and ensure the motor normal and stable and avoid the occurrence of some accidents.

Acknowledgements

This work was supported by the National Natural Science Foundation of China (No. 31300783), China Postdoctoral Science Foundation (No. 2014M561458), Doctoral Fund of the Ministry of Education Jointly Funded Project (No. 20123121120004), the Shanghai Maritime University Research Project (No. 20130474), the Shanghai Top Academic Discipline Project- management science & engineering, and the high-tech research and development program of China (No. 2013A2041106).

References

[1] Wang Z X 2008 Data Mining and Condition Recognition on Characteristic Data of Quayside Container *D. Shanghai Jiaotong University*

[2] Ma Y and Ruan Y 2011 *J.* Vector Quantization Method Based on Improved SOFM Neural Network *J. Journal of Computer Engineering and Science*

[3] Yi F Research on Extraction Method of Network Security Situation Elements Based on Evolutionary Neural Network Model *J. Electronic world.*

[4] Xiao K and Yuan S C 2010 Application of SOM Neural Network in Fault Diagnosis of Rotating Machines *J. Mechanical Design and Manufacturing.*

[5] Kohonen T 1998 The Self-organizing Map *J.*Neurocomputing.
[6] Kohonen T 1982 Self-organized formation of topologically correct feature maps *D. Biological Cybernetics.*
[7] Chen C P 2009 Crane control using maehine vision and wand following *J.IEEE International.*
[8] Zheng Z Y, Chen YY and Zhang J etal. 2013 An Improved Pre-processing Algorithm of Radar Signal Sorting Based on SOFM Clustering *J. Aerospace Electronic Warfare*
[9] Dai S B, Lei W H and Cheng Y Z etel. 2015 Research on Combined SOFM and SVC Clustering Method for Pre-deinterleaving of Radar Pulse *J. Firepower and Command Control.*
[10] Ivrissimtzis I P, Jeong W K and Seidel H P, 2003 Using growing cell structures for surface reconstruction. *Proceedings of the Shape Modeling International.*
[11] Dai S B, Lei W H and Cheng Y Z etal. 2014 Clustering of DOA data in radar pulse base Don SOFM and CDBW *J.Journal of Electronics.*

The Effect of the Khong Wong Yai Parameters on Sounds by FEM

Joompon Bamrungwong[1] and Pattaraweerin Woraratsoontorn[2]

1 Industrial Physics and Medical Instrumentation, Faculty of Applied Science, King Mongkut's University of Technology North Bangkok,1800 Bangkok, Thailand
2 Industrial Physics and Medical Instrumentation, Faculty of Applied Science, King Mongkut's University of Technology North Bangkok,1800 Bangkok, Thailand
E-mail: joompondej@gmail.com

Abstract. The Khong Wong Yai is called Gong in English but different in number to tune the sounds, beeswax with not exactly amount is pasted under the bossed. It is a Thai's percussion instrument usually involves some kind of striking on nipple gong. It is composed of 16 units of Gong as a circle of a rattan frame. The bossed or nipple gong embossed in center of a gong generates one key sound. Player has to percuss the gongs to create the rhythm with two hard rubber mallets or soft padded mallets as beaters and sits in the middle of the circle. The sound of the gong usually involves vibration of its own structure. The aim of this analysis is to study the eigen modes and find out the important parameters to generate sound frequency which are excited on the bossed or nipple gong by finite element analysis. In this paper the inertia relief method is to study the sound frequency of the Gong. The results show that the diameter of the Gong, the thickness of flange and the thickness of cylinder formed the Gong are important parameters which can be optimized to obtain the correct sound key.

1. Introduction

The melodies are very necessary to all forms of music of all cultures. As well known the percussion instrument is important to generate them. The percussion instrument often involves some kind of striking to the object and then makes vibration that produce the sound. Although there are so many percussion instruments for traditional Thai musical instruments, the khong wong yai is a type of the percussion instrument that is structured as the circle with the many gongs in a rattan frame. It is frequently occurred in the Piphat ensemble to provide the skeletal melody with the other instruments of the refined ensemble[1]. The bossed or nipple gong is a gong that is embossed in center. The sixteen key sounds are generated from the sixteen different sizes bossed gongs in a rattan frame. The gongs are individually tuned with beeswax under the circular metal disc. A player sits in the middle of the circle and holds two hard rubber mallets or soft padded mallets for hitting the gong for beating. However, the engineering principles were not being applied to design and produce the gongs by handmade, while they were constructed with the experience and skill of the maker. Bor-Tsuen Wang studied the Chinese's gong by finite element method and attempt to verify by experimental modal analysis (EMA)[2]. However, the theory of Euler-Bournoulli equation is often used for studying the vibration of the object that four parameters are needed for the solution of the equation for only fixed supports[3]. Actually, the gong is suspended vertically by a cord passed through two holes adjacent to the top rim of the bowl structure that can be assumed as free in the space not fixed support. Thus, it is impossible to determine the eigen modes of the gong by Euler-Bournoulli beam theory[4,5]. This research proposed to investigate the modes shape characteristics of the gong's sound with no constraint body using inertia relief by FEM. The simulation result achieved was validated by comparison with the experiment result.

2. Materials

Basically, an individually gongs is adjusted the key tone with beeswax pasted under the circular metal disc. In this study, the first and the last gongs comprised in the khong wong yai without beeswax thoroughly were analyzed to generate the most difference of frequency among the gongs. The first bossed gong is larger size than the last one which can generate the lower frequency than another. Although the general shape of both gongs are same, their sizes are different. Therefore, the eigen modes were validated by sound spectrum analyzer. However, the 3D model actual size the gongs were created for analysis by Ansys finite element software. The khong wong yai is shown Fig. 1(a). The first gong and the last gong used in this study is shown in Fig. 1(b).

In this investigation, both bossed gongs were made of the brass. The engineering data of material properties were used for FEA has density 8600 kg/m3, Young's modulus 1.06x105 MPa and Possion's ratio 0.31.

(a)

(b)

Figure 1. (a) The khong wong yai (b) The first gong (left) and the last gong (right)

3. Method

Due to the vibration of the Khong Wong Yai is a free vibration that can be considered with no constraint on it. In previous research[4] the inertia relief approach is combined with finite element analysis in the modeling and analysis of unconstrained systems. In this study, the commercial finite element packages ANSYS is used for analysis. The parameters have been studied is shown in Fig. 2.

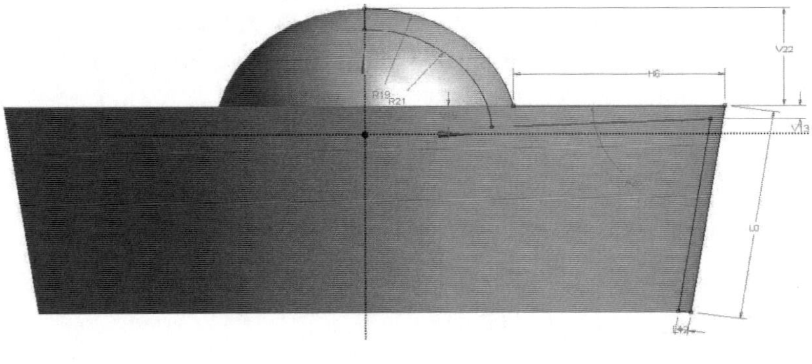

Figure 2. The parameters of the Khong Wong Yai are analyzed.

The governing equation most use to explain the transverse frequency is Euler-Bournoulli beam theory that start from The equations of motion of the wooden bar can be written down using Newton's second law of motion. Consider the Figure 2 the Gong is to the transverse direction of vibration that the deflection. The wooden bar no external force is applied and if EI(x) and A(x) are assumed to be constant, equation (1) simplifies so that free vibration is governed by

$$\frac{\partial^2 w(x,t)}{\partial t^2} + \frac{EI(x)}{\rho A(x)} \frac{\partial^4 w(x,t)}{\partial x^4} = 0 \qquad (1)$$

the general solution of equation (1) can be calculated to be of the form

$$X(x) = a_1 \sin(\beta x) + a_2 \cos(\beta x) + a_3 \sinh(\beta x) + a_4 \cosh(\beta x) \qquad (2)$$

Then the frequency of wooden bar in hertz can be written in the form

$$f_n = \frac{\beta_n^2}{2\pi} \sqrt{\frac{EI(x)}{\rho A(x)}} \qquad (3)$$

The sound wave propagating through an elastic medium define as speed of sound was equal to the square root of the elastic property divided by its density, and from definition of radius of gyration the equation (3) can be written in the form

$$f_n = \frac{\vartheta \beta_n^2 R(x)}{2\pi} \qquad (4)$$

Where

$$\vartheta = \sqrt{\frac{E}{\rho}}$$

$$R(x) = \sqrt{\frac{I(x)}{A(x)}}$$

Noted that the natural frequency is depend on constraint of the boundary conditions but this analysis the Gong is freeing vibrate can be considered no constrain on the Gong. The result from finite element will be plotted and linearized by find out the slope $\frac{\Delta y}{\Delta x}$ then get unit per effect.

4. Results

Eight parameters of the gong were studied by FEA simulation more than 3 points as shown in the Fig. 3 to Fig. 9. From analysis results found that parameters V13, V15, H6, and L12 have high unit per effect respectively as shown in Table 1.

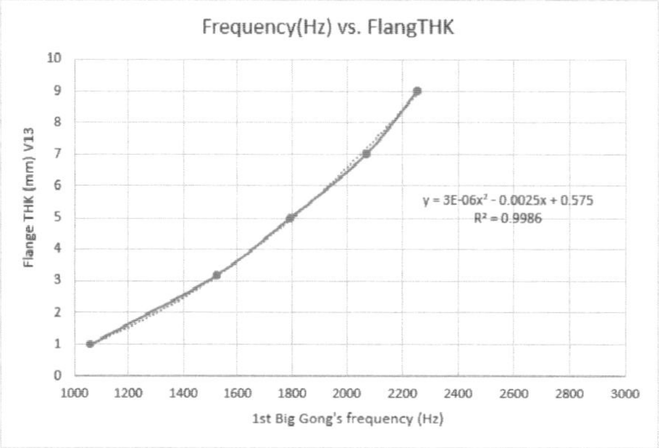

Figure 3. Frequency vs Flange thickness

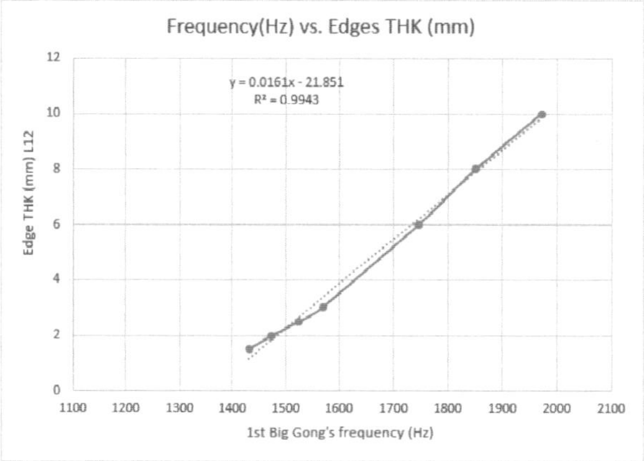

Figure 4. Frequency vs Edges thickness

IOP Conf. Series: Journal of Physics: Conf. Series **870** (2017) 012023 doi:10.1088/1742-6596/870/1/012023

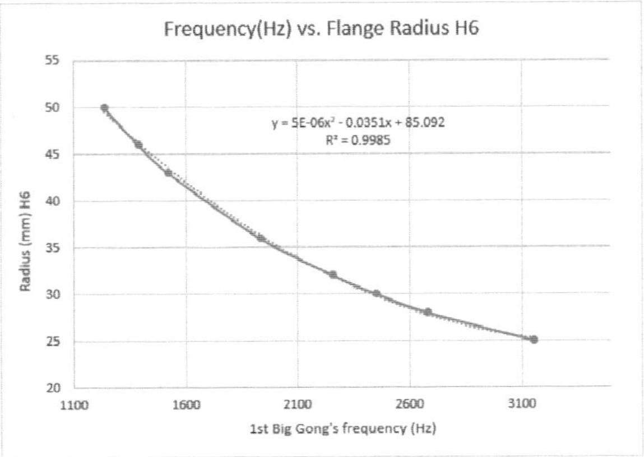

Figure 5. Frequency vs Flange radius

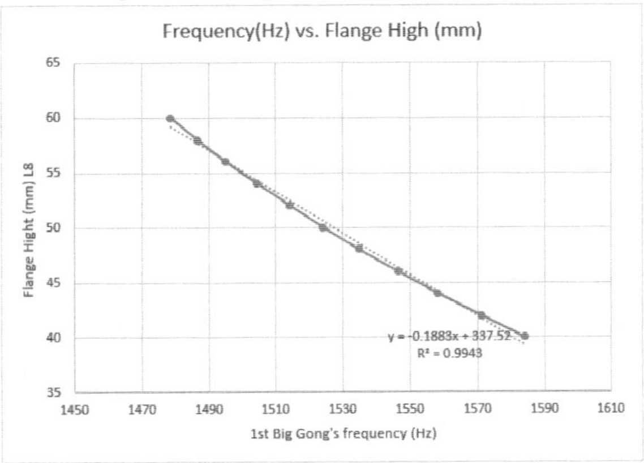

Figure 6. Frequency vs Flange high

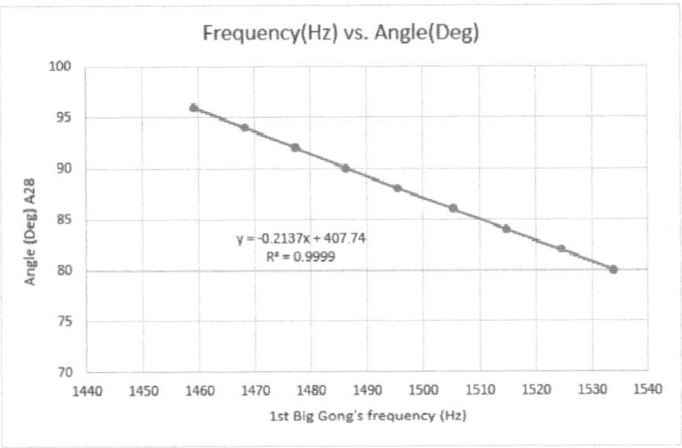

Figure 7. Frequency vs Flange angle

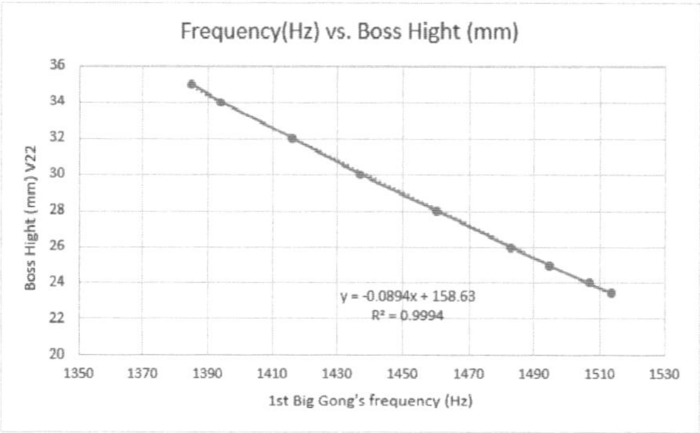

Figure 8. Frequency vs Boss high

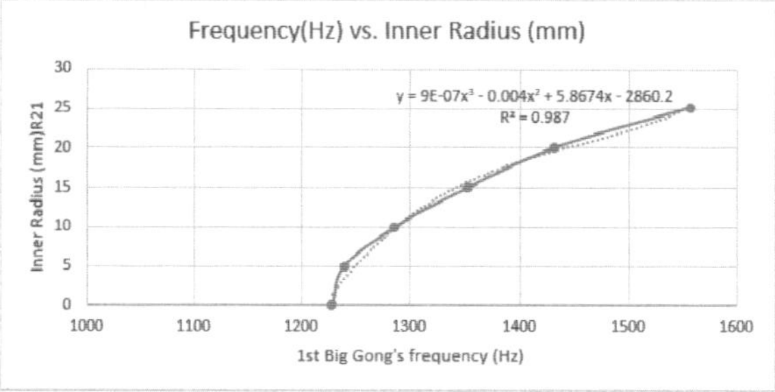

Figure 9. Frequency vs inner radius

Table 1. Unit per effect of Gong Wong Yai

Parameters	Remark	Unit	Unit/Effects
Thickness under boss_R21	Radius	mm	15
Boss high_V22	High	mm	12
Cylinder perspective_A28	Angle	Deg	5
High of cylinder_L8	High	mm	5
Dia. of flange_H6	Radius	mm	72
THK of cylinder_L12	Thickness	mm	68
THK of flange_V13	Thickness	mm	168
V15	Thickness	mm	146

5. Conclusion

The result show in Table 1 take a look at the Unit/Effect, from this initial study found that the thickness of the flange (v13) of the Khong Wong Yai are the key parameter influence and related to the tuning the sound frequency of the Khong Wong Yai. The secret of Khong Wong Yai is the thickness of its flange (v13 and v15). By FEA study the sound frequency of the khong wong yai has the eigen modes generated from bending mode that is moving upwards and downwards of the entire flange (top surface) of the bossed gong. The researcher recommend that drill the holes to hold the gong that it should be far a bit from the flange of embossed surface to avoid fast decay of the sound. In this study revealed that Khong Wong Yai needs beeswax for tuning the key sound always, The Ranat or the Xylophone can be designed with no the beeswax. Thus, Khong Wong Yai, It is impossible to make it without beeswax by all parameters are considered as linearity.

Reference

[1] Gong. [Online]. 30 May 2016, available: https://en.wikipedia.org/wiki/Gong
[2] Bor-Tsuen Wang. Intergation of FEA and EMA Techniques for Percussion Instrument Design Analysis. International Conference on System Science and Engineering, China (2011).
[3] J. Bamrungwong. Analysis and Design The Sound Frequency of a Treble Gamelan Using FEA. MENETT#20, Thailand (2006)
[4] J. Bamrungwong . The Sound Frequency Analysis of the Khong Wong Yai by FEM. International Conference Facilitating Autonomous Learning via Research-Based Approaches (FCAL), Thailand (2016)
[5] J. Bamrungwong, The Application of Inertia Relief Method in Ranat and Xylophone, International Conference on Electrical, Mechanical and Industrial Engineering, Thailand (2016)

Acknowledgement
I would like to thanks the department of Industrial Physics and Medical Instrumentation, Faculty of Applied Science, King Mongkut's University of Technology North Bangkok for highly support this research. And also thanks Ajan Amarin Wonsetti for helping proof this material.

IOP Conf. Series: Journal of Physics: Conf. Series **870** (2017) 012024 doi:10.1088/1742-6596/870/1/012024

Risk Assessment on Constructors during Over-water Riprap Based on Entropy Weight and FAHP

Tongqing Wu[1,2*] ,Liang Li[2*] ,Zelong Liang[3] , Tian Mao[2]and Weifeng Shao[2]

1 School of Civil Engineering and Architecture, Chongqing University of Science & Technology, Chongqing, 401331,China
2 National Engineering Research Center for Inland Waterway Regulation, Chongqing Jiaotong University, Chongqing 400074, China
3 Department of Hydraulic Engineering, Tongji University, Shanghai 200092,China
E-mail: 634986248@qq.com; wutongq@163.com

Abstract. Being aimed at waterway regulation engineering, there exist risks of over-water riprap for constructors which keeps uncertainty and complexity. For the purpose of evaluating the possibility and consequence, this paper utilizes fuzzy analytic hierarchy process with abbreviation of FAHP to do empowerment on the related risk indicators, constructs FAHP under entropy weight and establishes relevant evaluation factor set and evaluation language for constructors during over-water riprap construction process. Through doing risk probability estimation and risk consequence size evaluation on the factor of constructors, this paper introduces this model into risk analysis on constructors during over-water riprap of Ching River waterway regulation project. Results show that evaluation of this method is so credible that it could be utilized in practical engineering.

1. Introduction

Under the strategic guidance of Yangtze River economic zone, comprehensive transportation network has been constructed around Yangtze River which has already achieved initial success through doing integration on waterway, highway, railway and airline. However, much potential in Yangtze River shipping still remains to be fulfilled. It is particularly reflected in obstruction of upper and middle reaches of channel, which seriously hinders promotion of effective gathering and transferring system and improvement of proportion of high-level channel [1]. "Intestinal obstruction" of Jingjiang River channel is so serious that development of Yangtze River economic zone is greatly restricted. Therefore our government invested 4.3 billion yuan to dredge the "nine-bend and eighteen-curve" waterway of Jingjiang River. Meanwhile flood control is also taken into consideration whose main construction technologies include paving, rip-rapping, permeable frame flipping and slope protection with gabion mesh, etc. However, its construction quality may be influenced by multiple factors which keep various uncertainties. Therefore, it seems particularly important to make rational analysis on typical qualitative indexes.

Analytic hierarchy process is adopted for quantitative analysis on qualitative elements. However, the analysis needs frequent adjustment and consistency examination which increases calculation workload. Then FAHP is established to do improvement. Li Fawen[2] takes rural drinking water system as example and utilizes FAHP to do calculation on weights of dominant factors in brittle hierarchy structure thus evaluating safety of rural drinking water system from brittleness angle. Aiming at one tunnel collapse, Chen Jiejin [3] utilizes FAHP to construct one risk assessment model thus acquiring construction risk level of this section and comparing it with practical condition of construction site, whose conditions are identical. Jin Lianghai[4] constructs the comprehensive assessment model FAHP to do comprehensive evaluation and analysis on all involved indexes, which exempts deficiency of much adjustment and consistency examination of AHP and improves computational efficiency. The model is introduced in XiaoWan hydro-power project of Yunnan Province, whose reliability acquires verification. However, subjective factors of experts during its analysis process greatly influence the final results thus decreasing the credibility of analyzing results.

For the purpose of overcoming the subjectivity and uncertainty in the determination of risk assessment index proportion, entropy coefficient is introduced into fuzzy comprehensive evaluation to implement abundant research. For example, Gu Yujiong [5] introduces entropy weight, combines it with AHP and does scientific analysis on equipment maintenance mode of power station through considering objectification of experts' opinions. Researchers like Wu Dingyong [6] adopt matter-element entropy coefficient model to do safety evaluation on navigational environment of five shipping lanes. In a practical engineering project, researchers like Zhang Lingwan [7] compared the entropy coefficient fuzzy comprehensive evaluation with fuzzy comprehensive safety evaluation which verifies the rationality of model. In the assessment on flood risk degree of Chongqing-Guizhou expressway, researchers like Chen Hongkai adopted the method and got an accurate assessment. Luo Jungang [9] constructed entropy coefficient fuzzy comprehensive evaluation model and did evaluation analysis on risk of water shortage. Yang Yuzhong[10] adopted entropy coefficient fuzzy comprehensive evaluation method to assess and analyze safety of coal transportation.

Focusing on the complicity and uncertainty of risks for constructors in rip-raping of Jingjiang River Waterway regulation project, this paper adopts entropy coefficient method to analyze risk indexes of constructors in rip-raping of waterway regulation project, constructs FAHP under entropy coefficient and establishes relevant evaluation factor set and evaluation markers on constructors in rip-rapping to assess and analyze the risk level which accurately reflects risk level of constructors, makes assessment results keep reliability and has certain application value in physical engineering.

2. Method

2.1. *Fuzzy analytic hierarchy process*

Analytic Hierarchy Process (AHP) sets out all factors which are involved in analyses and decisions and divides them into levels of objective, principle, scheme, etc. On the basis of it, "qualitative analysis and quantitative analysis" could be realized. Decisions or assessment process can be quantified according to ideal and psychological rules. Since the method is especially pragmatic and effective for complicated assessment problems, it is widely used in risk assessment in various fields.

The key to AHP is the construction of assessment matrix. For sake of meeting requirements of adjustment and consistency examination, the matrix should be adjusted repeatedly, which is quite time-wasting and inefficient. If analysis keeps many layers and high complicity, it would be rather difficult to modify the judgment matrix. Therefore the idea of introduction of fuzzy math is utilized to determine assessment results during construction of judgment matrix. Generally speaking, there are reciprocal judgment matrix, complementary judgment matrix and mixed judgment matrix [11]. This paper adopts fuzzy complementary judgment matrix. That is, if matrix $A=\left(a_{ij}\right)_{m\times n}, i,j=1,2,\ldots n$, for any i,j, there are $a_{ij}=0.5, i=1,2,\ldots,n, 0<a_{ij}<1, a_{ij}+a_{ji}=1$; for matrix $A=\left(a_{ij}\right)_{m\times n}$, if there is a $k\left(1<k<n\right)$, then $a_{ij}=a_{ik}-a_{jk}+0.5$, it is a fuzzy complementary judgment matrix.

2.2. *Entropy and entropy coefficient*

In information science, there exists information entropy which presents the uncertainty or the probability an object may occur. It is a measurement of the disorder degree of a system. From the practical use of a piece of information, the size of entropy stands for the amount of information. If the information amount is very large, then the event will be of great certainty and the entropy value will

IOP Conf. Series: Journal of Physics: Conf. Series **870** (2017) 012024 doi:10.1088/1742-6596/870/1/012024

be small; on the contrary, if the information amount is quite small, then the event will be uncertain and the entropy value will be big.

Suppose that the system is under multiple conditions and the probability varies in each condition $p_i\left(i=1,2,\ldots,m\right)$, then the entropy of the system can be defined as $H=-k\sum_{i=1}^{m}p_i\ln p_i$, in the formula, there are m conditions to be assessed and n indexes already assessed. Then a matrix of $m\times n$ is constructed. That is, if $Q=\left(q_{ij}\right)_{m\times n}$,for any q_{ij} the information entropy can be:

$$H_j=-k\sum_{i=1}^{m}p_{ij}\ln p_{ij} \tag{1}$$

In the above-mentioned formula, p_{ij} is the normalization of q_{ij}.That is $p_{ij}=\frac{q_{ij}}{\sum_{i=1}^{m}q_{ij}}$, k is the coefficient and there is $k=\frac{1}{\ln n}$; if $p_{ij}=0$, then $H_j=0$.The entropy coefficient of j index can be defined as:

$$w_H^j=\frac{1-H_j}{\sum_{j=1}^{n}\left(1-H_j\right)} \tag{2}$$

From above, the priority vector of assessment index should be: $W_H=\left(w_H^1,w_H^2,\cdots,w_H^n\right)$.Here, the entropy coefficient stands for the fierce degree of competition among all indexes.

3. Construction of Fuzzy Analytic Hierarchy Process Assessment Model-Entropy Coefficient

For sake of lowering the risks, hydro geological conditions, raw materials, shipping equipment and construction site during rip-raping process of waterway realignment are all strictly controlled. However, it is complicated and difficult to control worker element in the process. As it could not be ignored, it would be necessary to make risk assessment for constructors. Therefore, AHP is adopted and fuzzy math is introduced to "obfuscate" the judgment matrix. After that a fuzzy comprehensive judgment matrix is constructed and entropy coefficient is utilized to make the result be more objective.

3.1. Index factor set of risk assessment for constructors in rip-raping

According to relevant materials and field research, risk carriers in rip-raping can be divided into six types including raw materials, shipping equipment, hydro geological conditions, constructors, environment inside and outside the construction site. This paper focuses on the risk assessment on constructors in rip-raping and makes a risk factor set U, which analyses the probability and consequences of risks for constructors in the construction. It is shown in figure 1.

3.2. Risk assessment markers

Risk assessment on workers in rip-raping is aimed at highlighting constructors. The probability and consequences of the risks are put into five grades. For the probability, there are extremely low, low, medium, high and extremely high. The formula is $v_1=\left\{v_{11},v_{12},v_{13},v_{14},v_{15}\right\}$ and the degree increases from v_{11} to v_{15}; for the consequences, there are extremely low, low, medium, serious and extremely serious. The formula is $v_1=\left\{v_{21},v_{22},v_{23},v_{24},v_{25}\right\}$ and the degree increases from v_{21} to v_{25}.

3.3. Calculation and judgment on overall weight

In the entropy coefficient comprehensive assessment model, there are subjective weight in FAHP and objective entropy coefficient in fuzzy comprehensive assessment as well as entropy coefficient process.

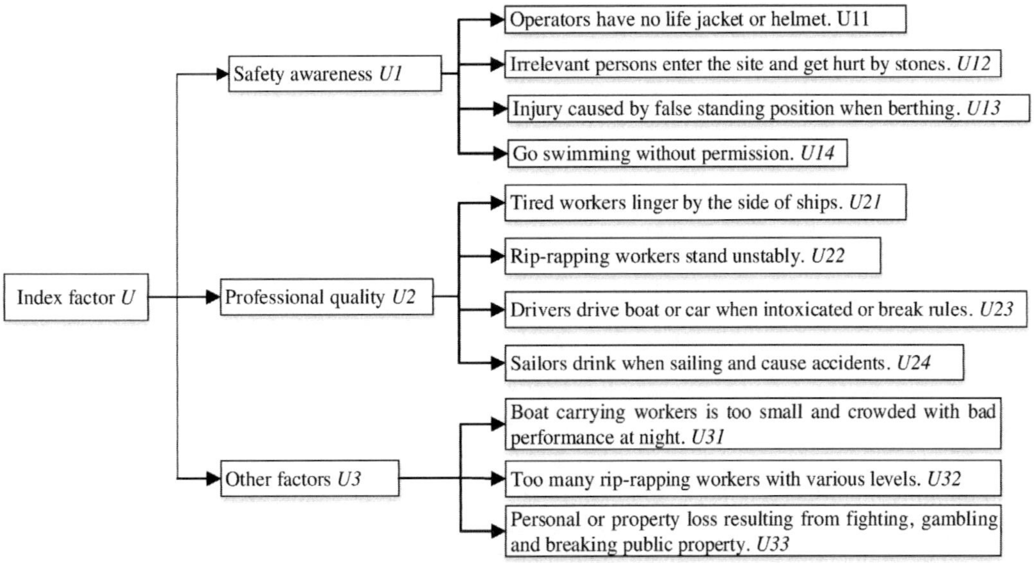

Figure 1 Structure of index factors of constructors

3.3.1 Subjective and objective weight. If $U=\{u_1,u_2,\ldots,u_n\}$ is the assemblage of n factors on assessment object, then the judgment matrix is $U^{(k)}=\begin{bmatrix} r_{11} & r_{12} & \cdots & r_{1n} \\ r_{21} & r_{22} & \cdots & r_{2n} \\ \vdots & \vdots & & \vdots \\ r_{n1} & r_{n2} & \cdots & r_{nn} \end{bmatrix}$, The rule for the value of r_{ij} is

$r_{ij}=\begin{cases} 0.5 & t(i)=t(j) \\ 1 & t(i)>t(j) i,j=1,2,\cdots,n \\ 0 & t(i)<t(j) \end{cases}$, $t(i)$ and $t(j)$ stands for the relative importance of factor $u(i)$ and factor $u(j)$ respectively.

Fuzzy consistency judgment matrix is transferred through r: $R=\begin{bmatrix} R_{11} & R_{12} & \cdots & R_{1n} \\ R_{21} & R_{22} & \cdots & R_{2n} \\ \vdots & \vdots & & \vdots \\ R_{n1} & R_{n2} & \cdots & R_{nn} \end{bmatrix}$, In the formula,

$R_{ij}=\frac{k_i-k_j}{2n}+0.5$, $k_i=\sum_{j=1}^{n} r_{ij}$, $i=1,2,\ldots,n$."Addition" is adopted to calculate relative weight. That is subjective weight and the result is $A=\{a_1,a_2,\ldots,a_n\}$; the ideas of all experts are integrated into a fuzzy comprehensive assessment matrix. Based on the system and the calculation procedure in 1.2, the objective weight is $W=\{w_1,w_2,\ldots,w_n\}$.

3.3.2 Overall weight. From above, we can know that subjective weight is $A=\{a_1,a_2,\ldots,a_n\}$ and objective weight is $W=\{w_1,w_2,\ldots,w_n\}$.Then the overall weight of i index is: $t_i=\frac{a_i \times w_i}{\sum_{i=1}^{n} a_i \times w_i}$, Vector of the overall weight: $T=\{t_1,t_2,\ldots,t_n\}$ (3).

3.3.3 Overall assessment result. For single-level result of overall assessment, fuzzy comprehensive evaluation is adopted based on formula: $B_i=T_i^{\circ} R_i$ (4).

IOP Conf. Series: Journal of Physics: Conf. Series **870** (2017) 012024 doi:10.1088/1742-6596/870/1/012024

That is: $(b_1,b_2,...,b_n)=(t_1,t_2,...,t_n)^{\circ}\begin{bmatrix} r_{11} & r_{12} & \cdots & r_{1n} \\ r_{21} & r_{22} & \cdots & r_{2n} \\ \vdots & \vdots & & \vdots \\ r_{n1} & r_{n2} & \cdots & r_{nn} \end{bmatrix}$ .In the formula B_i stands for single-level comprehensive assessment result; R_i stands for fuzzy comprehensive evaluation matrix; T_i stands for weight vector; "o" stands for fuzzy operator.

Among all sub-assessment factors, all formulas above can work out their fuzzy comprehensive judgments whose results are $B_1, B_2,..., B_m$. Therefore, the fuzzy comprehensive evaluation matrix of the factor assemblage is: $B=\begin{bmatrix} B_1 \\ B_2 \\ \vdots \\ B_m \end{bmatrix}=\begin{bmatrix} b_{11} & b_{12} & \cdots & b_{1n} \\ b_{21} & b_{22} & \cdots & b_{2n} \\ \vdots & \vdots & & \vdots \\ b_{m1} & b_{m2} & \cdots & b_{mn} \end{bmatrix}$ .So forth, the fuzzy comprehensive evaluation result of the top layer is B. This study focuses on the importance of a factor and processes the fuzzy evaluation result with the maximum subordination principle. That is: $V=\left\{V_L \middle| V_L \to \max_{j=1} b_j\right\}$.

4. Applications of Living Examples

4.1. Determination of Assessment indexes

In order to issue management document of strong pertinence to prevent risks and guide workers, factors influencing the safety of workers must be collected and analyzed through entropy coefficient-FAHP. All risk factor indexes can be seen in figure 1 whose assessment markers are in 2.2.

Ideas of experts on the probability and consequences of the risks are collected to form the second layer of fuzzy comprehensive assessment model. B_1 stands for possibility and B_2 stands for consequences. .B_{1i} and B_{2i} represent the assessment matrix:

$$Q_{11}=\begin{bmatrix} 0.16 & 0.24 & 0.34 & 0.2 & 0.06 \\ 0.12 & 0.12 & 0.32 & 0.32 & 0.12 \\ 0.10 & 0.2 & 0.32 & 0.2 & 0.18 \\ 0.12 & 0.3 & 0.28 & 0.2 & 0.10 \end{bmatrix},\ Q_{12}=\begin{bmatrix} 0.18 & 0.2 & 0.32 & 0.24 & 0.06 \\ 0.18 & 0.2 & 0.32 & 0.24 & 0.06 \\ 0.2 & 0.28 & 0.2 & 0.28 & 0.04 \\ 0.08 & 0.2 & 0.24 & 0.38 & 0.1 \end{bmatrix}$$

$$Q_{13}=\begin{bmatrix} 0.12 & 0.16 & 0.3 & 0.34 & 0.08 \\ 0.16 & 0.24 & 0.24 & 0.26 & 0.1 \\ 0.14 & 0.22 & 0.3 & 0.26 & 0.08 \end{bmatrix},\ Q_{21}=\begin{bmatrix} 0.12 & 0.24 & 0.22 & 0.18 & 0.24 \\ 0.16 & 0.3 & 0.34 & 0.14 & 0.06 \\ 0.1 & 0.26 & 0.28 & 0.26 & 0.1 \\ 0.16 & 0.18 & 0.24 & 0.26 & 0.16 \end{bmatrix}$$

$$Q_{22}=\begin{bmatrix} 0.08 & 0.24 & 0.24 & 0.34 & 0.1 \\ 0.10 & 0.36 & 0.22 & 0.2 & 0.12 \\ 0.10 & 0.26 & 0.24 & 0.26 & 0.14 \\ 0.14 & 0.22 & 0.32 & 0.26 & 0.06 \end{bmatrix},\ Q_{23}=\begin{bmatrix} 0.02 & 0.2 & 0.22 & 0.38 & 0.18 \\ 0.04 & 0.2 & 0.36 & 0.32 & 0.08 \\ 0.08 & 0.26 & 0.22 & 0.34 & 0.1 \end{bmatrix}$$

Based on the assessment matrix Q_{1i}、$Q_{2i}\ (i=1,2,3)$ above and formula (1) and (2), the entropy coefficient is calculated: $W_1^1=(0.31,0.31,0.16,0.22)$; $W_1^2=(0.19,0.22,0.27,0.32)$; $W_1^3=(0.49,0.19,0.32)$; $W_2^1=(0.10,0.50,0.32,0.08)$; $W_2^2=(0.31,0.25,0.15,0.29)$; $W_2^3=(0.36,0.41,0.23)$.

Assessment matrix in FAHP model is:

Assessment matrix of the first layer: $P=\begin{bmatrix}0.5 & 1 & 1\\ 0 & 0.5 & 1\\ 0 & 0 & 0.5\end{bmatrix}$

Assessment matrix of the second layer:

$$P_1=\begin{bmatrix}0.5 & 0.5 & 1 & 0.5\\ 0.5 & 0.5 & 1 & 0.5\\ 0 & 0 & 0.5 & 0\\ 0.5 & 0.5 & 1 & 0.5\end{bmatrix},\ P_2=\begin{bmatrix}0.5 & 1 & 0.5 & 0\\ 0 & 0.5 & 0 & 0\\ 0.5 & 1 & 0.5 & 0.5\\ 1 & 1 & 0.5 & 0.5\end{bmatrix},\ P_3=\begin{bmatrix}0.5 & 1 & 0.5\\ 0 & 0.5 & 0\\ 0.5 & 1 & 0.5\end{bmatrix}$$

All the weight coefficients are worked out through "addition": the first layer$\rightarrow$ the second layer: $A=(0.454,0.333,0.213)$; The second layer$\rightarrow$the third layer: $A_1=(0.283,0.283,0.151,0.283)$, $A_2=(0.25,0.151,0.283,0.316)$, $A_3=(0.392,0.216,0.392)$.

Based on formula (3), the overall weight of the second layer is $T_1^1=(0.335,0.335,0.092,0.238)$, $T_1^2=(0.184,0.128,0.296,0.392)$, $T_1^3=(0.536,0.114,0.350)$, $T_2^1=(0.118,0.588,0.201,0.094)$, $T_2^2=(0.311,0.151,0.17,0.368)$, $T_2^3=(0.441,0.277,0.282)$.

Last, overall assessment can be worked out through formula (4) and the Fuzzy Comprehensive Assessment matrix in the second layer:

$B_{11}=T_{11}{}^{\circ}R_{11}=(0.132,0.210,0.317,0.240,0.101)$, $B_{12}=T_{12}{}^{\circ}R_{12}=(0.147,0.224,0.253,0.307,0.070)$,

$B_{13}=T_{13}{}^{\circ}R_{13}=(0.132,0.190,0.293,0.303,0.082)$, $B_{21}=T_{21}{}^{\circ}R_{21}=(0.143,0.274,0.305,0.180,0.099)$,

$B_{22}=T_{22}{}^{\circ}R_{22}=(0.109,0.254,0.266,0.276,0.095)$, $B_{23}=T_{23}{}^{\circ}R_{23}=(0.043,0.217,0.259,0.352,0.130)$,

The fuzzy comprehensive assessment matrix of the first layer is:

$$R_1=\begin{bmatrix}0.132 & 0.210 & 0.317 & 0.240 & 0.101\\ 0.147 & 0.224 & 0.253 & 0.307 & 0.070\\ 0.132 & 0.190 & 0.293 & 0.303 & 0.082\end{bmatrix},\ R_2=\begin{bmatrix}0.143 & 0.274 & 0.305 & 0.180 & 0.099\\ 0.109 & 0.254 & 0.266 & 0.276 & 0.095\\ 0.043 & 0.217 & 0.259 & 0.352 & 0.130\end{bmatrix}$$

It is seen from the matrix above that overall weight of the first layer $T_1=(0.280,0.353,0.367)$ and $T_2=(0.236,0.272,0.492)$. By $B_1=T_1{}^{\circ}R_1=(0.137,0.210,0.286,287,0.083)$, $B_2=T_2{}^{\circ}R_2==(0.085,0.241,$ $0.272,0.291,0.113)$ and the maximum subordination principle, which tells that the probability of risks is "high" and the consequence of risks is "serous". The results agree with the actual situations. The workers are easily influenced by the factors above. The risks will cause severe loss in both aspects of property and personal security, which shows that assessment result is quite reliable.

4.2. MATLAB programming

To make it more easily for construction managers to control the project and to simplify the calculation process, *MATLAB* is adopted to build in all the formulas above and *GUI* programming is used to make the interface more friendly and clear for operators to get risk grade.

5. Conclusions

Aiming at complicity and uncertainty of risks on workers in rip-raping during Jingjiang River waterway regulation process, this paper adopts FAHP to analyze the risk indexes of the workers in rip-raping. Entropy coefficient method is also introduced during fuzzy evaluation to objectively deal with experts' ideas and constructs entropy coefficient-FAHP model.

Based on entropy coefficient fuzzy comprehensive assessment model, this paper constructs relevant assessment factor set and assessment markers for the risks of workers in rip-raping. Through doing risk probability estimation and risk consequence size evaluation, this paper does programming MATLAB and introduces it into physical engineering, which is in accordance with field investigation. It shows strong adaptability of this model in assessment on Jingjiang River waterway regulation, which also keeps simple and convenient calculation operation. This would provide handy, scientific and reliable thoughts for other similar construction processes during the waterway regulation project.

6. Acknowledgments

This project is supported by National Engineering Research Center for Inland Waterway Regulation & Key Laboratory of Hydraulic and Waterway Engineering of the Ministry of Education of Chongqing Jiaotong University (SLK2014B02), Fundamental and Advanced Research Projects of Chongqing (cstc2015jcyjA30003), Research Foundation of Chongqing University of Science & Technology (CK2015Z29),Chongqing Municipal Commission of Urban-rural Development (2015-1-39),Scientific and Technological Research Program of Chongqing Municipal Education Commission (KJ1601336).

References

[1] Information on http://leaders.people.com.cn/n/2014/0925/c58278-25733540.html,2014-09-25.
[2] Li Fawen,Li Manman,Fu Changfeng.Research on the brittle source for the breakdown of the drinking water safety system[J].China Rural Water and Hydropower, 2012,(8):92~95.
[3] Chen Jiejin,Zhou Feng,Yang Jun-sheng,et al.Fuzzy analytic hierarchy process for risk evaluation of collapse during construction of mountain tunnel[J]. Rock and Soil Mechanics, 2009, 30(8): 2365~2370.
[4] Jin Lianghai,Wei Xiongwei,Li Huafei, et al.FAHP model and its application in post evaluation of hydropower project[J]. Yangtze River,2012, 43(1):95~98.
[5] GU Yujiong,Chen Kunliang,Yang Kun.Decision of maintenance mode for power plant equipment based on entropy weight and analytic hierarchy process[J]. Journal of North China Electric Power University,2008, 35(6):72~76+82.
[6] Wu Ding, Wen Yuanqiao. Risk evaluation of waterway environment based on entropy weight and matter element model[J].Journal of Wuhan university of technology (transportation science & engineering),2014, 38(5):1158~ 1162.
[7] ZHang Lingwan,Ding Hongfei, Chen Yanjin. Study on early warning for railway transport of dangerous goods based on entropy weight fuzzy comprehensive evaluation [J]. China Safety Science Journal,2012, 22(5): 119~125.
[8] Zeng Rong, Chen Hongkai,Li Junye.Fuzzy comprehensive evaluation method based on entropy in road flood risk assessment[J]. Journal of Chongqing Jiaotong university (natural science),2010,29(4):587~591.
[9] Luo Jungang,Xie Jiancang,Ruan Benqing. Fuzzy compreh -ensive assessment model for water shortage risk based on entropy weight[J].SHui Li Xue Bao,2008,39(9): 1092 -1097.
[10] Yang Yuzhong,Wu Liyun, Cong Jianchun.Fuzzy synthetic evaluation on safety of coal mine transportation systems based on entropy weight[J].Journal of Harbin Institute of Technology,2009,41(4):257~259.
[11] Xu Zeshui, Liu Haifeng.On compatibility of six kinds of uncertain judgement matrices[J]. Fuzzy Systems and Mathematics, 2003,17(2):53~58.

Calculation of Appropriate Minimum Size of Isolation Rooms based on Questionnaire Survey of Experts and Analysis on Conditions of Isolation Room Use

An-Na WON[1], Hae-Eun SONG[2], Young-Kwon YANG[3], Jin-Chul PARK[3] and Jung-Ha HWANG[1]

1 School of Architecture, Kyungpook National University, Daegu, Republic of Korea
2 School of Architectural Civil, Environmental and Energy, Kyungpook National University, Daegu, Republic of Korea
3 Department of Architecture, Chungang University, Seoul, Republic of Korea
E-mail: woan749@hanmail.net

Abstract. After the outbreak of the MERS (Middle East Respiratory Syndrome) epidemic, issues were raised regarding response capabilities of medical institutions, including the lack of isolation rooms at hospitals. Since then, the government of Korea has been revising regulations to enforce medical laws in order to expand the operation of isolation rooms and to strengthen standards regarding their mandatory installation at hospitals. Among general and tertiary hospitals in Korea, a total of 159 are estimated to be required to install isolation rooms to meet minimum standards. For the purpose of contributing to hospital construction plans in the future, this study conducted a questionnaire survey of experts and analysed the environment and devices necessary in isolation rooms, to determine their appropriate minimum size to treat patients. The result of the analysis is as follows: First, isolation rooms at hospitals are required to have a minimum 3,300mm minor axis and a minimum 5,000mm major axis for the isolation room itself, and a minimum 1,800mm minor axis for the antechamber where personal protective equipment is donned and removed. Second, the 15 ㎡-or-larger standard for the floor area of isolation rooms will have to be reviewed and standards for the minimum width of isolation rooms will have to be established.

1. Introduction

The MERS epidemic that spread to Korea in 2015 resulted in considerable damage, including 16,752 patients placed in isolation, 186 patients diagnosed with MERS, and 38 deaths[1]. After the outbreak, issues were raised over secondary infections caused within hospitals, hospitals' capabilities to manage infected patients, and the appropriate treatment of infected people. The most serious issue was the lack of isolation rooms essential for treatment of respiratory infections[2]. Existing isolation rooms did not have antechambers or multiple patients were placed in one isolation room. As health laws are likely to be strengthened in the wake of MERS, so these rooms will need improvements to comply with strengthened standards.

The Korean government is planning to revise medical regulations to substantially strengthen standards for isolation rooms at hospitals[2] : one or more for a general hospital with 300 or more beds and an additional one for every 100 more beds. Considering that it is difficult to immediately improve existing isolation rooms, the government is encouraging step-by-step improvement. The government has specified that an isolation room should have a floor area of at least 15 m2 and a separate antechamber. A total of 159 hospitals, including 116 general hospitals (41% out of total 281) and 43 tertiary hospitals (total 43 tertiary hospitals with 500 or more beds) will now have to meet standards for the required isolation rooms.

This study conducted a questionnaire survey of experts who have experience in managing hospital facilities and a survey of medical staff that have treated patients in isolation rooms in order to draw suggestions for establishing standards of the size of isolation rooms.

2. Method

Please Along with the questionnaire survey of experts, a survey was also conducted on the size of equipment kept in isolation rooms and the environment of their use, in order to determine the minimum size necessary for isolation rooms.

The questionnaire survey was conducted of 27 experts on hospital ventilation systems. Basic information about those experts is shown in Figure 1. Experts in their thirties were 25% of the total; forties, 63%; and fifties or older, 13%. They have an average of 16 years of experience in the field.

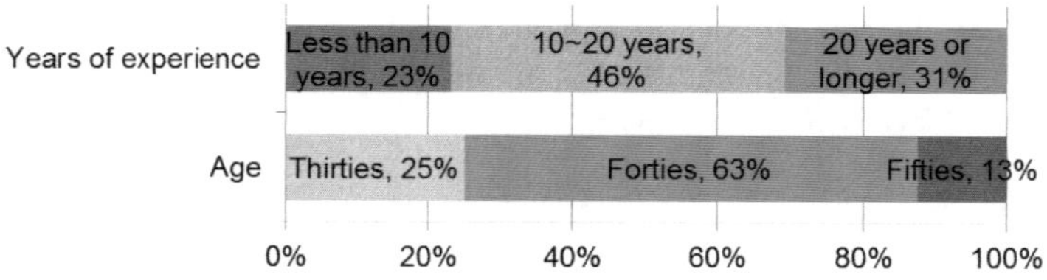

Figure1. Experts' age and years of experience

3. Research

3.1. Estimation of minimum size of isolation rooms based on the result of questionnaire survey on experts

The minimum total size suggested by experts of an isolation room and its front room are on average 28 m² and 7 m², respectively, which is slightly larger than the standard for government-designated hospitals (15 m²) as shown in Figure 2. Considering the number of staff needed when treating patients and that the number of movable medical appliances is larger for isolation rooms than for general wards, the size of new isolation rooms and their front rooms will have to be reconsidered.

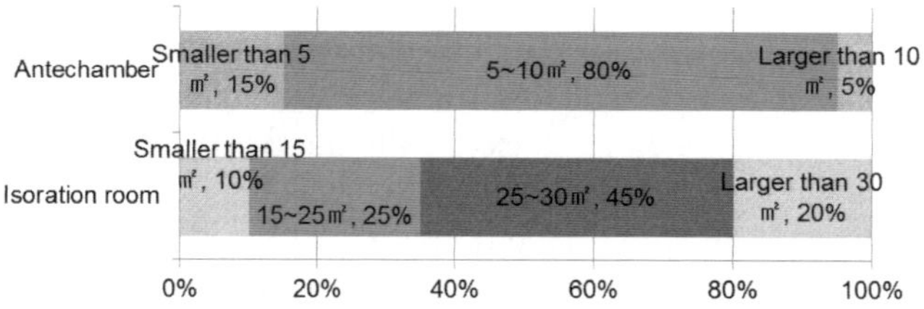

Figure 2. Size of isolation room and front room suggested by hospital ventilation system experts

3.2. Estimation of the size of isolation rooms accounting for the number of devices and staff needed for treatment

The severity of a patient's symptoms is classified as in Table 1. The manual[3] specifies that when there are not enough isolation rooms in a hospital, patients with mild or moderate-level symptoms be treated in a temporary isolation room and patients with severe symptoms (such as those that require ventilator treatment) undergo mandatory isolation in an isolation room. Therefore, respirators and medical appliances used in an ICU, such as electrocardiograms (ECG) and cardioverter defibrillators, are required equipment in an isolation room.

Table 1. Classification of severity of symptom and necessary care unit

Severity	Stage	Definition	Treatment Unit
Mild	1	No symptom	General care unit (Temporary isolation room)
	2	Symptom (fever, cough, dyspnoea, hemoptysis, nausea, abdominal pain, diarrhea, etc.)	
Moderate	3	Symptom of pneumonia but no oxygen demand	General care unit (Temporary isolation room)
Severe	4	Symptom of pneumonia and oxygen demand	ICU (Temporary isolation room)
	5	Need for airway intubation and ventilator treatment	Isolation room
	6	Application of ECMO	

The essential equipment needed in an isolation room can be classified as a bed, trollies, and medical appliances. The size of a bed [4][a] is approximately 2160*1050 mm and it has to be located near the center of an isolation room. The size of medical trollies [4][b] used for storing and carrying first-aid equipment is approximately 750*470 mm on average. Trollies are used for storing medicine, holding medical devices, and attaching integrated monitors, and two to three trollies are necessary for one isolation room.

As mentioned above, multiple types of medical appliances as well as trollies are necessary in an isolation room, because of the characteristics of respiratory infections. Notably, the appliances in an isolation room are banned from being moved in and out of the room during treatment in order to prevent the spread of the pathogen; therefore, they need to be stored within the room when not in use. The size and usage of medical appliances essential for an isolation room are in Table 2. Other than those appliances, IV stand poles and reserve linens are kept in an isolation room.

Table 2. Medical appliances needed for an isolation room and their size

Medical appliances	Usage	Size	Note
Electrocardiogram	ICU	Ranges from 320*280 mm to 410*320 mm (Need to consider the size of the stand)	Portable ones are available. These may be places on a stand at the top of the bed
Ventilator	Breathing assistance		Embedded (No separate space necessary)
Defibrillator	AED		Portable, so trollies can be used as the stand pole
ECMO	Extracorporeal membrane oxygenation	800mm of minimum width (Diverse according to its shape)	Used for a patient whose lungs or heart is failing

[a] There are different sizes of beds, but generally, they are manufactured between the minimum size of 2,000*900 mm and maximum size of 2,200*1,600 mm.

[b] Medical trollies are classified into ones with a storage space below the bed (930*650 mm, 673*480 mm, etc.) and light-weight trollies (630*470 mm, 539*366 mm, etc.). Their average size is estimated to be 750*470 mm.

Personal protective equipment (PPE) refers to clothes or devices that are worn in order to protect people from infectious materials[5]. Medical staff treating patients with respiratory infections is required to wear PPE, which largely consists of gloves, gown or aprons, face protection, and respiratory protection.

Table3. Type of PPE

	Components	Note
Face, eyes, and respiratory protection)	Surgical cap (head protection), Medical mask, Respirator, Goggles	The contaminated area must not contact the body or other equipment when putting on or taking off PPE
Skin and clothes protection	Scrubs, Apron, Overalls, Double gloves	
Shoes protection	Boots	

(N95 mask, gloves, safety goggles, face shield, gown)

Koreans between 30 and 34 years old [6] have on average a shoulder width of 432mm for men and 384mm for women, and a chest size of 990mm for men and 863mm for women. The space occupied when a person wears PPE over general clothes is estimated to be 1.2 to 1.4 times larger than the average body size. The principle of using PPE is to put it on before contacting a patient and to take it off immediately after patient care, right out of the isolation room. It is important that PPE be taken off in the antechamber. Thus, in an isolation room 518 to 604mm and 460 to 537mm of width for movement must be guaranteed for men and women, respectively. In the antechamber, where PPE is removed, the length of the arms[c] must be reflected in the width of the room, which needs to be at least 1800mm. Of course, a space with a width larger than the minimum will be needed in order to prevent contaminants from being smeared onto the walls of the antechamber.

The treatment carried out in an isolation room is classified into general treatment such as blood collection and medication, and emergency treatment for dyspnoea or seizure. The minimum staff necessary is one or two[7][d] for general treatment and two to four for emergency treatment.

4. Results

4.1. Calculation of isolation room's minimum size in consideration of physical elements

In consideration of the bed used in the room, installation of the bed and other medical appliances, and the number of patients treated in the room, the minimum width for an isolation room at a hospital is 1,050 mm for the bed; 604 mm for the shoulder width of medical personnel (sides of bed); and 800 mm for movable appliances such as trollies. The simple sum of those widths is 3,090mm, but when the space needed for movement is considered, at least 3,300mm width is required. In addition, the front room where PPE is donned and removed will need to have at least 1,800 mm width. This 1,800 mm is the width that is actually available for the movement of people and appliances and excludes the width of fixed medical appliances at the side of the room, as in the case when it contains a fixed storage of medical appliances.

For the depth (the major axis) of an isolation room, at least 4,650mm of width is needed, including 2,000mm for the length of bed, 1,000mm for the space for the installation of ECMO, and 604mm aisle width for the movement of two staff. When there is a fixed medical device or storage, the total width will have to be calculated excluding the width of this equipment. In such a case, a minimum of 1.2 times minimum room width will need to be applied as a reserve space, thus bringing the minimum width of the major axis of an isolation room to an estimated 5,000mm.

[c] The average arm length for adult men was 585 mm (aged between 30 and 34) and 533 mm for adult women, as of 2010.

[d] Two nurses, one cleaner, and one assistant for dressing and undressing, according to the guidelines for infectious disease management institutions on the working-level responses to MERS ver1.3

IOP Conf. Series: Journal of Physics: Conf. Series **870** (2017) 012025 doi:10.1088/1742-6596/870/1/012025

The current standard for the size of an isolation room is 15 ㎡. However, when the necessary minimum width calculated by this study is used to set the major axis and minor axis at 3,300 mm and 5,000 mm, the minimum floor area needed for an isolation room is 16.5 ㎡. In addition, the criteria for the minimum width of an isolation room will have to be established in further detail.

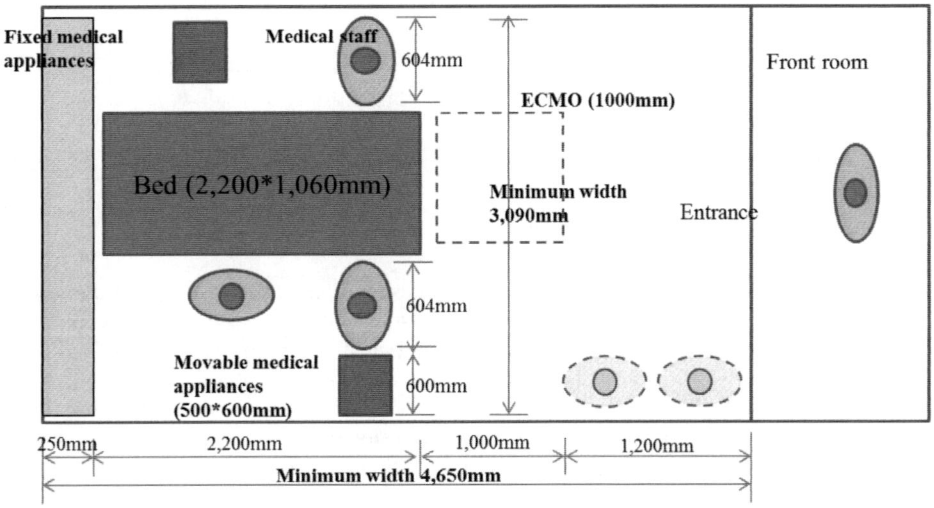

Figure 3. Size of an isolation room

5. Conclusion

Because infectious respiratory diseases such as SARS and MERS break out frequently, the importance of isolation rooms in hospitals to treat respiratory diseases and prevent their further spread is increasing. This study conducted a questionnaire survey of experts and analyzed the use of isolation rooms and their necessary appliances. This analysis determined the minimum room size needed for patient treatment in order to contribute to hospital construction plans for the future.

6. References

[1] KCDC, The 2015 MERS outbreak in the republic of Korea, pp185, 2016.07.
[2] Ministry of Health and Welfare, Press release (201.07.27) http://www.mohw.go.kr/front_new/al/sal0301vw.jsp?PAR_MENU_ID=04&MENU_ID=0403&page=1&CONT_SEQ=333614 (in Korean)
[3] KCDC, Guidelines for infectious disease management institutions on the working-level responses to MERS ver1.3, 2015.06
[4] Medical device manufacturer, http://korean.electric-hospitalbed.com (cited 01 Feb. 2017)
[5] The korean society of infectious diseases, http://www.ksid.or.kr/mail/file/MERS01.pdf, 2016.06.
[6] Korean Agency for Technology and Standards, The 6th project of three dimensions form measurement of Korean's body size, https://sizekorea.kats.go.kr/06_civilCenter/libraryView.asp?idx=4890&page=1, pp100-112, 2010
[7] MERS central branch medical institution, Guidelines for infectious disease management institutions on the working-level responses to MERS, 2015.06.

Acknowledgments
This work was supported by the National Research Foundation of Korea(NRF) grant funded by the Korea government(MSIP) (NRF-2016R1A2B2011255).

Research on fluorescence detection method of Microcystis aeruginosa

Xiao-xiong Wang

School of Civil and Environmental Engineering, Shenzhen Polytechnic, Shenzhen 518055, China
Email: wangxiaoxiong20@szpt.edu.cn

Abstract: The paper studied the viability determination of *Microcystis aeruginosa* by FDA and PI staining. The staining results were measured by fluorescence microscopy. The results indicated that viable and dead cells were stained as bright green and red fluorescent respectively by FDA and PI. Through PI-FDA dual color fluorescence staining, the color of green and red distinct obviously by fluorescence microscope. The staining rate has relation with the cell density. If the cell density of *M. aeruginosa* was 1.0×10^7-1.0×10^9 cell·mL^{-1}, the staining rate would be 100.0% or 98.0% by PI and of FDA respectively.

1. Introduction

In recent years a large number of industrial wastewater and domestic sewage had been discharged into, accelerate the process of eutrophication of lakes and reservoirs in the area of diffusion, year by year, duration prolongation. Taihu, Dianchi, Chaohu, Hongze lake has a "bloom", even the flow of the river, the largest tributary of the Yangtze River has also appeared as "algae bloom", algae problem exists in the world of water treatment practice[1].

Blooms of algae in water body not only causes the deterioration of water quality, the processing and safety of drinking water also have been influenced, threaten the safety of drinking water: some algae releases caused by algae toxins zoo noses or even death[2,3], algae cells and its extracellular secretions produced three trihalomethanes, haloacetic acids, halogenated ethylene cyanide substances such as in the chloride process[4-7]. Due to algal blooms, it poses a serious threat to the safety of drinking water for human beings, attracted great attention of researchers at home and abroad. The removal and inactivation of algae become one of the urgent problems to be solved.

The establishment of quantitative detection method to measure the algae removal effect is need for algae research, the detection method is simple and efficient will contribute to the in-depth study. Usually use the algae cell dense chemical method, algae cell density decreased the number is not large, large number of algal cells still exist in water, whether these cells with activity, culture can be determined only through again, and again cultured long time[8, 9]. There are also in water treatment, using biomass index to indicate the algicidal effect, but the index does not accurately reflect the algicidal effect, by measuring the changes of dissolved oxygen in the water containing algae, indirectly judge the survival state of algal cells [10], the method also need to cultivate algae, consumption time is longer, the discriminant but cannot directly cell activity state [11].

This research is based on the PI and FDA fluorescent dye single staining method, detection of *Microcystis aeruginosa* activity of PI-FDA double staining, were observed and statistical result of staining by fluorescence microscopy, in order to provide a basis and reference for

IOP Conf. Series: Journal of Physics: Conf. Series **870** (2017) 012026 doi:10.1088/1742-6596/870/1/012026

the identification of algal cell activity in water treatment.

2. Materirals and methods

2.1. Culture of M. aeruginosa

M. aeruginosa test using was purchased from Chinese Academy of Sciences, Wuhan Institute of hydrobiology. Algae is a freshwater lakes and rivers in common algae. With BG-11 culture medium were cultured in 5 L white fine mouth bottle, a temperature of 26±0.5 ℃, the light intensity of 30 μmol·m^{-2} s^{-1}, light dark ratio 12h:12h, shaking 1~2 times a day.

2.2 Method and equipment

2.2.1 Reagent

PI (propidium iodide, Sigma), FDA (fluorescein diacetate, Sigma).

2.2.2 Instrument and equipment

Fluorescence microscopy: OLYMPUS BX51; centrifuge HC3018 (Zhongke Zhongjia); ultra pure water system of MILLIPORE-V3.05.

2.2.3 Preparation of cell suspension

The addition of 1 ml Lugeshi solution at 10 ml of algal cells in liquid, in darkness for 48 h, the cells were fixed. Fixed cells lost all activity, but the individual cells were not broken decomposition. Centrifugal liquid fixed algal cells after 3000 r·min^{-1} speed, 3~4 times of culture medium suspension with, until Lugeshi solution was clean and the solution became colorless. Reconstitution of cell death was as cells staining.

2.2.4 Preparation and dyeing dyes

FDA is soluble in acetone, match into solution concentration of 5 mg·ml^{-1}. 4 ℃ preservation refrigerator, staining the concentration of 100 μg·ml^{-1}.

PI DPBS buffer solution (Dulbecco 's phosphate buffered saline) with concentration of 400 g·ml^{-1} solution, keep in the refrigerator 4 ℃, staining the concentration of 60 μg·ml^{-1}.

PI single staining: take the algae cell suspension 1 mL, add 12.5 μL PI to 1mL algal cell suspension, 5 min reaction.

FDA single staining: take the algae cell suspension 1mL, 2 μLFDA, evades the light reaction of 5 min. PI/FDA double staining: algae cell suspension was 1 mL, 2 μLFDA and 12.5μLPI, evades the light reaction of 5 min.

2.2.5 Fluorescence microscopy

Take in the algal cell liquid logarithmic growth period, prepared into solution, containing different density of live cells join FDA, placed 5 min at room temperature the darkness. Take the algae cell liquid after fixation, using DPBS to prepare solution, containing different density of dead cells entering PI, but also placed 5 min at room temperature.

Take the living and dead cells in mixed cell suspension, to join the FDA, the oscillation mixing after joining PI, placed 5 min. Fluorescence microscopy of FDA detection, the maximum excitation wavelength of 493 nm, PI maximum excitation wavelength of 540 nm. In the blue light, living cells by FDA dyed bright was green. Dead cells were PI stained was red, staining with camera.

2.2.6 Cell count

Algal cell density by 0.1 mL phytoplankton counting plate count, the formula for

IOP Conf. Series: Journal of Physics: Conf. Series **870** (2017) 012026 doi:10.1088/1742-6596/870/1/012026

calculating the:

$$N = \frac{A}{\mathrm{B} \times \mathrm{C}} \times \frac{D}{E} \times F$$

The formula for the algal cells per liter of water in the number; for counting frame area (mm^2); as a vision of the area (mm^2); to count view number (a); 1 L water concentration (mL); volume after volume counting frame(mL) for each piece; measured by the number of algae.

On FDA and PI single stained cells, according to the number of cells under the bright field and fluorescent cells, under the field quantity, calculating the efficiency of dyeing cells.

3. Results and discussion

3.1 Algal cell morphology observation

Algae without staining, the results were shown in Figure 1, the figure shows that the algal cell is composed of a single cell or double cell individual algae mostly spherical, untreated *Microcystis* in vesicular, comparative morphology has rules, in accordance with relevant literature reports.

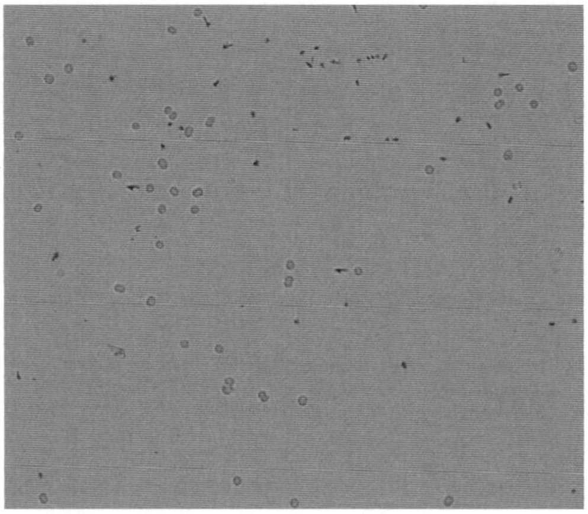

Fig.1 Micrographs of *M. aeruginosa*' single-cell unstained in bright field（200×）

3.2 PI Fluorescent staining

PI staining results in Figure 2, the PI cannot enter cells, when cell apoptosis or death, can enter the cell through the cell membrane, and intracellular DNA and RNA phase function generation red fluorescent material, dead cells emit red fluorescence, as can be seen from the graph, PI at 630 nm excitation light green, dying cells were dyed red, bright red fluorescence emitted. Research shows that, the algal cell membrane integrity refused to enter the PI dying, cell membrane damaged cells to enter within the PI and combined with the nucleic acid, so the fluorescence intensity of PI was stronger that cell membrane damage is more serious in this [12-14], and research related to the previous agreement.

Fig.2 Fluorescence micrographs of M. aeruginosa' single-cell stained by PI（200×）

3.3 FDA fluorescent staining

FDA penetratethe algae cells, esterase hydrolysis produced by transformation of fluorescein, the excitation/emission wavelength of 495nm/525 nm, the fluorescence intensity is proportional to the activity of esterase, therefore using fluorescence intensity characterization of esterase activity of fluorescein. FDA is a non fluorescent hydrophobicderivatives of fluorescein, can penetrate the cell membrane into the cell, and catalyzes the hydrolysis of two acetate groups, has a high intensity of fluorescence of fluorescein in the product, under blue light excitation (495 nm), live cells were dyed bright green FDA, cell membrane damaged, FDA can not be accumulated in the cells, cell death could not emit fluorescence. In this experiment, FDA fluorescence staining is shown in Figure 3, it is shown that the fluorescent substance full of whole cells, cells emit bright green fluorescence, according to, can not only calculate the number of cells, and can clearly distinguish the cell shape.

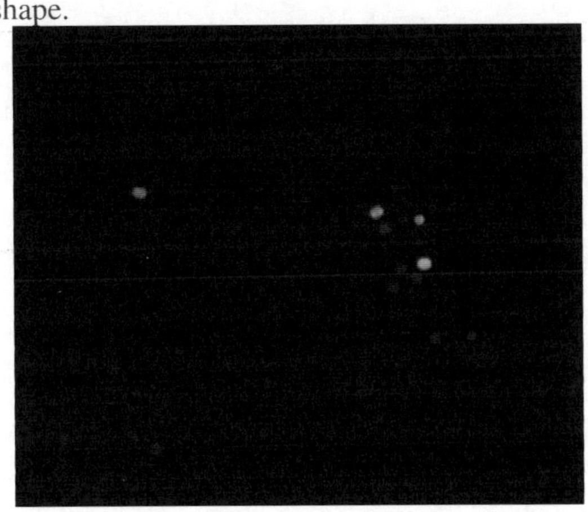

Fig.3 Fluorescence micrographs of *M. aeruginosa*' single-cell stained by FDA（200×）

3.4 FDA-PI double fluorescent staining

The 1× 107 L^{-1}, M. aeruginosa cells after fixation, respectively and live cell solution according to 1:1 into cell suspension, live cells accounted for 50%, accounted for 50% of death cells, Stained by FDA-PI double fluorescence, shown in Figure 4, it can be seen from the figure, live cells were dyed bright green, and death cells were stained red. Under the fluorescence microscope, green and red distinction is clear, can be judged on cell activity, green for living cells, and red for the death of cells.

Fig.4 Fluorescence micrographs of *M. aeruginosa*' single-cell stained by FDA--PI（200×）

3.5 Effects of M. aeruginosa cell density of staining efficiency on FDA and PI

FDA and PI on the growth of *M. aeruginosa* single staining efficiency results are shown in Table 1, it can be seen from the table, the cell density has little effect on the efficiency of dyeing, in $1x10^7$~$1.0x10^9$ cell mL^{-1} range, PI staining efficiency higher than FDA, PI is relatively easy to enter cells, red light emitted by the fluorescent cell death comparison, clearly visible, and the long holding time, the dying efficiency can reach above 98%.

Table 1 The staining rate of FDA and P I to *M. aeruginosa*

Cell density（cell·mL^{-1}）	1.0×10^7	5.0×10^7	1.0×10^8	1.0×10^9
Staining efficiency of FDA/%	92.0	94.0	97.0	98.0
Staining efficiency of PI /%	98.0	100.0	100.0	100.0

4. Conclusions

FDA hydrolysis of fluorescein accumulation in cells and live cells emitted green fluorescence under blue light, the green fluorescent cells in living cell. While the PI is just the opposite, in blue light, the death of P I cells were dyed red, red fluorescence for dead cells. Therefore, it was feasible to detect algae cell activity using PI and FDA monochromatic fluorescence.

From the experimental results, PI is relatively easy to enter cells, red fluorescent cell death sends, comparison is clearly visible, and keep for a long time, the dyeing efficiency can reach above 98%.

In the range of $1x10^7$~$1.0x10^9$ cell mL^{-1}, the cell density has little effect on the efficiency of dyeing, cell concentration is larger, the efficiency of dyeing is higher, but the difference is not very obvious.

5. References

[1] Wang X X, Jiang C C and Zhu J 2011 Progress in the measuring methods of algae cell viability. *Environmental Engineering* **29**: 248-248
[2] Lahtl K, Rapala J, Kivimaki A L, et al. 2001 Occurrence of microcystins in raw water sources and treated drinking water of finnish waterworks. *Water science technology*, **43(12)**: 225-228
[3] Beasley V R 1989 Algae intoxication in livestock and water flow [J]. *Vet Clin North Am Food Anim Pract* **5(2)**:345-361
[4] Graham N J D, Wardlaw V E, Perry R, et al. 1998 The significance of algae as trihalomethane precursors [J]. *Water science technology* **37(2)**: 83-89
[5] Hoehn R C, Barnes D B, Thompson B C, *et al.* 1980 Algae as sources of trihalomethane precursors [J]. *J Am Water Works Assoc* **72(6)**:344-350
[6] Oliver B G and Shindle D B 1983 Trihalomethanes from the chlorination of aquatic algae [J]. *Environmental science technology* **17(2):** 80-83
[7] Wang L, Fang J Y, Ma J, *et al.* 2005 Effect of pre-oxidation on algal cell morphology and algal removal by enhanced coagulation process [J]. *Journal of southeast university (Natural science*

edition) **35(sup I)**: 182-185

[8] Canavate J P and Lubińn L M 1995 Some aspects on the cryop-reservation of microalgae used as food for marine species. *Aquaculture* **136(3)**: 277-290

[9] Liang W Y，Wang K，Ruan L L, *et al.* 2010 Study on the variation of algal activity during the electrochemical oxidation as inactivation method. *Environmental scienc*e **31 (6):** 14642-1469

[10] Wu X W, Gao T Y and Li G J 2000 New electrochemical technique of water treatment disinfection and algae killing. Acta scientiae *circumstantiae* **20 (sup.)**: 75-79

[11] Chen L B, Liang W Y, Qu J H, *et al.* 2005 The viability determination of cyanobacteria by double staining with fluorescein diacetate propidium iodide. *Environmental chemistry* **24(5)**: 554-557

[12] Yu Y, Kong F X, Wang M L, *et al.* 2007 Determination of short-term copper toxicity in a multispecies microalgal population using flow cytometry. E*cotoxicology and Environmental Safety* **66(1)**: 49-56

[13] Zhang M, Kong F X, Wu X D, *et al.* 2008 Different photochemical responses of phytoplankters from the large shallow Taihu Lake of subtropical China in relation to light and mixing. *Hydrobiologia* **603(1)**: 267-278

[14] Su W, Kong F X, Yu Y, *et al.* 2013 Effects of the rice straw on *Microcystis aeruginosa* analyzed by different physiological parameters . *Environmental science* **34(1)**: 150-155

Acknowledgements

This work was supported by the National Natural Science Foundation (NSFC) of China, Grant No. 5137 8316 and the Science, Industry, Trade and Information Technology Commission of Shenzhen Municipality (JCYJ20130331151242230) for their financial support of the research studies described in this paper.